AF867579

FROM LOGGERS TO FORESTERS

FROM LOGGERS TO FORESTERS

The Newfoundland Overseas Forestry Unit (1939–1946): Wartime Labour, Colonial Politics, and Postwar Recognition

URSULA A. KELLY

MEMORIAL UNIVERSITY PRESS

Library and Archives Canada Cataloguing in Publication
Title: From loggers to foresters : the Newfoundland Overseas Forestry Unit (1939–1946) : wartime labour, colonial politics, and postwar recognition / Ursula A. Kelly.
Other titles: Newfoundland Overseas Forestry Unit (1939–1946) : wartime labour, colonial politics, and postwar recognition
Names: Kelly, Ursula A. (Ursula Anne), 1956- author
Description: Includes bibliographical references and index.
Identifiers: Canadiana (print) 20260164038 | Canadiana (ebook) 20260164089 | ISBN 9781990445576 (softcover) | ISBN 9781990445583 (EPUB) | ISBN 9781990445590 (PDF)
Subjects: LCSH: Newfoundland Overseas Forestry Unit—History. | LCSH: World War, 1939-1945—Regimental histories—Great Britain. | LCSH: World War, 1939-1945—War work—Great Britain. | LCSH: World War, 1939-1945—War work—Newfoundland and Labrador. | LCSH: Lumbermen—Newfoundland and Labrador—History—20th century.
Classification: LCC D768.154.N49 K45 2026 | DDC 940.53/718—dc23

Copy editing: Sandy Newton
Cover image: The Loading - Landings, Newfoundland Lumberjacks at Work in Scottish Forest by David Macbeth Sutherland, courtesy of the Aberdeen Art Gallery
Cover design and page layout: Alison Carr

Published by Memorial University Press
Memorial University of Newfoundland and Labrador
P.O. Box 4200, St. John's, NL A1C 5S7
www.memorialuniversitypress.ca

Printed in Canada

31 30 29 28 27 26 1 2 3 4 5 6 7 8

Funded by the Government of Canada | Financé par le gouvernement du Canada | Canada

CERTIFIED CANADIAN PUBLISHER

DEDICATION

To the memory of my parents—Margaret J. Waterman, teacher, originally of Torbay, and Andrew J. Kelly, logging camp foreman and contractor, originally of Burnt Island, Bonavista Bay—whose life together began at Gambo in the months following the end of the Second World War.

CONTENTS

Section Two

Section Three

Conclusion

ILLUSTRATIONS

Images

Tables

ACKNOWLEDGEMENTS

While researching this manuscript, I benefitted from the expertise of archivists in both Canada and the United Kingdom, including those at The Rooms Provincial Archives (St. John's, NL), Archives and Special Collections at Memorial University, The National Archives (London, UK), and the Aberdeen City and Aberdeenshire Archives (Scotland). I extend a special thank you to Don Walsh, who oversees the Digital Archives Initiative of the Queen Elizabeth II Library at Memorial University.

I am grateful to the skilled staff of Memorial University Press. In her role as interim academic editor, Dr. Vicki Hallett reviewed the initial manuscript draft. I am thankful for her input and support. I am delighted to have worked once again with Alison Carr, managing editor and book designer, and Randy Drover, book publicist. I am especially grateful to academic editor and exceptional scholar Dr. Fiona Polack, who provided support, guidance, and consideration at a crucial time.

I am pleased that editor Sandy Newton agreed to prepare the manuscript for publication. Sandy is an extraordinary editor whom I trust and appreciate. I am also thankful to Alexandra Peace for her excellent work in preparing the index.

Several years ago, while completing research for *The Foresters' Scribe*, a book about the First World War Newfoundland Forestry Companies, Dr. Melvin Baker suggested that I follow it with a book about the Newfoundland Overseas Forestry Unit of the Second World War. Melvin is an accomplished historian of Newfoundland and Labrador who is also known for his scholarly generosity and support of others and their work. I am very grateful to be one of the many who have benefitted from his collegiality.

Dr. Peter Neary's historical writings about Newfoundland related to the period under consideration in this book were essential to my research. As one of a very few historians who have written about the Newfoundland Overseas Forestry Unit, Peter understood how its story was tossed in the swell created by the monumental shifts occurring in Newfoundland in the 1930s and 1940s. He also saw how the efforts of its members for postwar recognition were consequential for the wider arc of Canadian citizenship. Peter had graciously agreed to write a foreword for this book, but died in early 2024, several months before I finished the manuscript draft. I will always gratefully remember his enthusiastic support of this project.

I acknowledge, with thanks, that the research for this book was supported by a grant attached to a John Lewis Paton Distinguished University Professorship at Memorial University. The book was published with the help of a grant from the Federation for the Humanities and Social Sciences, through the Scholarly Book Awards, using funds provided by the Social Sciences and Humanities Research Council of Canada. I am also grateful for the assistance of a grant from the J. R. Smallwood Foundation for Newfoundland and Labrador Studies.

This manuscript emerged in difficult personal times. It was written over several years in the solitude and quiet of early mornings, a space carved by the urge to write despite the constraints of daily life. My love of writing was born in childhood under the tutelage of my mother, with whom I spent countless hours scribbling at the kitchen or living room tables of our family home in

Gambo, Newfoundland, while she did household chores around me. Those times with her inscribed writing and study as comfort and security.

Numerous teachers encouraged my writing, but none more or with greater influence than Lisa de Leon, my remarkably talented high school English teacher and friend. In recent years, I have become deeply thankful to Dr. Stacy M. Smith, clinical psychologist, who has helped me in many ways, including by recognizing and affirming my need to write despite limitations, which freed me to do so without compunction.

Finally, to my friends Rhonda and Clar, and to family members who, in various and sometimes extraordinary ways and often despite great geographic distance—a special acknowledgement of Yolanda, MJ, Ed, Andrea, and Ron in these regards—have reached out, stepped up, or held fast: your care and love carry me, and I am ever grateful.

ABBREVIATIONS

ANDCO – Anglo-Newfoundland Development Company; AND Company
NCG – Newfoundland Commission of Government
NOFA – Newfoundland Overseas Foresters' Association
NOFU – Newfoundland Overseas Forestry Unit
UK – United Kingdom of Great Britain and Northern Ireland
VAC – Department of Veterans Affairs, Government of Canada
(now Veterans Affairs Canada)

INTRODUCTION

THE NEWFOUNDLAND OVERSEAS FORESTRY UNIT (1939–1946): WARTIME LABOUR, COLONIAL POLITICS, AND POSTWAR RECOGNITION

Timber was an essential material for the Allies in the Second World War. It was crucial for making pit props, the support beams used to build the mines from which coal—the industrial energy source for war—was excavated. The tonnage of timber required by the British Expeditionary Force exceeded that of any other must-have item (Lloyd, 1945). Huge quantities of timber were required to construct and maintain trenches; to build barracks, duckboards, and fence posts; to make poles for communication lines and anti-landing structures; and to build and repair bridges, railways, ships, and even airplanes.[1]

The country had also experienced an extraordinary demand for timber just a few decades earlier. At the time of the First World War, the forest industry of the United Kingdom (UK) had been quite small. The nation imported more than 90 percent of its timber supply from its colonies, the Scandinavian countries, and Russia (Oosthoek, 2013). The Great War compromised access to this supply and disrupted its transportation routes, rendering the UK vulnerable not only in its wartime needs but potentially again in the future. This risk, in turn, made developing a home supply a priority after the war.

Despite lessons learned during the First World War about the risks of overreliance on imported timber and the efforts of the UK Forestry Commission to develop a state-owned supply, the country was still relying almost

1 Thousands of the de Havilland Mosquito combat planes were built during the Second World War in the UK, Canada, and Australia. The plane—sometimes called the Wooden Wonder—was noteworthy for its wooden construction and, as a result, for its lightness and speed.

exclusively on imports by the start of the Second World War. The time between the wars had been insufficient for growing a mature home supply. As during the First World War, importing timber required large amounts of shipboard cargo capacity—tonnage—which, during wartime, was also needed for other essentials. German U-boat warfare threatened the reliable supply of all imported essential goods, including timber.

When the Second World War began, a huge local supply of timber was not all that was needed. Harvesters, sawyers, and lumberers to cut and mill timber for specific uses were also scarce, as the existing UK labourers were not trained in woods work. Many workers who were not in "reserved occupations"—positions in which they were legally required to remain in wartime—were at the war front. Caught once again in a dire situation regarding timber supply and related labour, the UK looked to its dominions, including Newfoundland, for help. In response, in late 1939, only a few short months after the onset of war, the first contingent of Newfoundland woods workers left for the forests of the UK.[2]

Over the war's span, these workers, members of the Newfoundland Forestry Unit, would number approximately 3,600. The unit was one of the many ways the Newfoundland Commission of Government (NCG), a form of unelected government in place in Newfoundland at the time, responded to the needs of the Allies. Commonly known as the Newfoundland Overseas Forestry Unit (and thus, as the NOFU), it was a civilian entity—a workforce not an armed force—and the largest dedicated group of temporary workers to leave the country in support of the UK war effort. It was largely settler-constituted[3]

2 Throughout this book, I use the official name for what is now the Canadian province of Newfoundland and Labrador that is appropriate to the historical period under discussion: "Newfoundland," the official name of a country until 1949 and of the province from 1949 to 2001; thereafter "Newfoundland and Labrador." The two main geographical regions of the province are referred to as Labrador and the Island (i.e., Newfoundland). I use the term "Newfoundlanders" to describe only those who inhabit the island of Newfoundland and "Labradorians" to describe those who inhabit Labrador; when referring to both groups, I use "Newfoundlanders and Labradorians." Indigenous groups—Mi'kmaq, Innu, and Inuit—are named specifically where known and appropriate

3 Settlers are those people and their descendants who migrated from Europe to what is now Canada during the colonial era, displacing, sometimes violently, the original or Indigenous Peoples and establishing socio-cultural and political dominance.

and included men from hundreds of primarily anglophone and several francophone rural communities on the Island. It also included Mi'kmaw men and a few men from Labrador. The NOFU's contribution to wartime timber supply in the UK was, in the end, substantial, but there is much more to its story, including how the unit's formation, wartime experiences, and postwar struggles were affected by the historical and political context of Newfoundland in the first half of the twentieth century.

The unit's impact was also personal and inter-generational, reverberating in families and communities on both sides of the Atlantic. In one of his memoirs, Scottish writer Ian Cameron wrote of the transformation of his home by the NOFU, who built camps near his hometown of Ballater and whose work he observed as a young boy. He described the unit's time in Scotland as "one minor migration that made of my boyhood world a new-found land for Newfoundlanders" (Cameron, 2008, p. v). This migration resonated most deeply in the social and cultural life of the Highlands of Scotland, where the thousands of NOFU men worked alongside other foreign nationals and locals; it was also felt in Newfoundland, to where, in the postwar period, hundreds of Scottish women immigrated with their NOFU husbands.

Whether its members remained in Scotland or returned to Newfoundland, which was by then at a political and cultural watershed (the decision to return to the self-government that had been suspended in 1934 or join the Canadian Confederation as a tenth province), the struggles and experiences of the unit while overseas and at home after the war reveal much about this period in the history of Newfoundland and Labrador. It is a story that is still largely untold.

Existing Accounts

With the exception of *Timber!!: Newfoundland Overseas Forestry Unit,* a souvenir booklet about the unit's time in Scotland that was published in 1945 by the Newfoundland Overseas Foresters' Association (NOFA), existing accounts of the work of the NOFU are few and not widely accessible. The best known, *They Also Served,* is a short account of the unit's work written by one of its superintendents, Thomas Curran, that offers an expanded account of the material included in *Timber!!*.[4] A small number of memoirs contain short chapters on member experiences with the unit (Gladney, 1999; Jeddore, 2015). A collection of short recollections by women who married men of the NOFU provides details of life in Scotland and immigration to Newfoundland following the war (Barrett et al., 1996). There are also a few locally published accounts of the NOFU that provide some details about men from specific areas or regions (Green, 2023; Hyslop, 2001; Martin, 1994). Together, these accounts provide a sketch of the work of the NOFU and glimpses of life beyond the workday. Ian Cameron's writings also provide insight into the socio-cultural impact of the NOFU in Scotland, particularly in Ballater, Aberdeenshire, near where some of the men worked between 1940 and 1943 (Cameron, 2008, 2010).[5]

The NOFU is also largely absent from published accounts of other Second World War colonial forestry units. It is not mentioned at all in some of them or, in a slight of a different order, is subsumed under Canada's contribution to the war effort.[6] Still other accounts under-represent the extent of its contribution, for example by providing grossly inaccurate numbers of enlisted workers.[7] These historical omissions and errors are unfortunate in light of the extent of the NOFU contribution and given Newfoundland's status at the time—it was not a province of Canada but a separate dominion. Although one of only two civilian overseas wartime forestry units (the other being the British Honduras Forestry Unit)—all other dominion forestry units were military—the NOFU was a very large and productive supplier of wartime timber and lumber and was the longest-serving of any of the forestry units, military or civilian.

The NOFU also rarely factors into war-based scholarship. Peter Neary's intermittent discussion of the NOFU in his book *Newfoundland in the North Atlantic World, 1926–1949* (1988), and his account of postwar efforts for

4 Some of the material in Curran is reproduced, along with a nominal roll of workers and their assigned numbers, on a website dedicated to the NOFU: www.nofu.ca.

5 For an instalment of Cameron's memoir, see Kelly & Forsyth (2018).

6 For examples, see G. M. Long (1952), McKelvey (2001), and Vickers (2011).

7 See, for example, Lloyd (1944).

recognition (1993b) are exceptional instances of scholarship that present an analysis of the politics of the unit. There is very little scholarship arising from the UK context, as well. Noteworthy are an archaeological study related to site remains at the NOFU camp in Strathmashie forest at Macoul, near Laggan (Sneddon, 2007), and a comparative architectural study of a Scottish prisoner-of-war camp and a NOFU woods camp as forms of institutional accommodation during the Second World War (Banks, 2011).[8] Finally, there is a short comparative study of certain aspects of the experiences of the British Honduras Forestry Unit and the NOFU that is framed around issues of colonialism and race (Forbes, 2015).

Alongside this dearth of published material, interest in the NOFU remains high in both the UK and among Newfoundlanders and Labradorians, and especially among descendants of unit members, many of whom participate in online communities where they share images and stories and attempt to fill gaps about their ancestors' service.[9] In an age of accessible global travel, descendants also travel between the UK, especially Scotland, and Newfoundland and Labrador seeking to resolve questions of family genealogy that are both historical and personal.[10]

In communities in Scotland, especially those in the north where the majority of NOFU camps were located and where residents' lives and landscapes were transformed by the woods workers, the NOFU impact, along with that of the other forestry units, is still felt—not only as a wartime memory, but also in how family, community, and landscape are understood. Faced with so many migrations as a result of the thousands of wartime marriages between

---

8 While there is very little that remains of any of the UK camps, some replication efforts are underway, the most significant being at Glenmuick camp, Aberdeenshire, by the Ballater Historic Forestry Project Association, a registered charity in Scotland.

9 See, for example, the group webpage entitled Newfoundland Overseas Forestry Unit, which in late 2025 had 2,100 subscribers: https://www.facebook.com/groups/Nnofu/.

10 One of the larger recorded pilgrimages is that of the Power family of Scotland. In summer 2018, a group of twenty-one people representing three generations of descendants of James Leo Power (NOFU #2341) of Broad Cove (now Duntara) travelled from Scotland to Newfoundland to visit his home and meet their Newfoundland relatives. See Squibb (2018). Power married in Scotland and remained there after the war.

local women and overseas personnel, hundreds of which included NOFU members, plus the addition of new immigrants—those men who arrived during the war and remained—ways of thinking about family and culture changed to reflect this transnationalism.[11] As a result, there is an abiding interest in the unit and its times.

Purpose

To fill in some of the blanks, I sought primary sources in archives in both Canada and the UK, which provided additional detail and texture. I also reviewed unit and government reports and correspondence, parliamentary transcripts, and newspaper and magazine articles in order to more fully piece together the story of the NOFU. Archival materials, while incomplete, nonetheless offer greater insight into government and unit processes, politics, and perspectives; newspapers and magazines capture how the public's reception of the men was informed and shaped, while also filling in some details unavailable in official sources or where official sources themselves are unavailable. Together, these historical sources provide a more in-depth picture of the unit, from its formation to the postwar period.

My purpose in writing this book through the use of these materials is twofold: first and foremost, to map more fully the experiences of the unit over the sixty years from its emergence at the start of the Second World War to the recognition of its service under the Government of Canada Pension Act at the end of the twentieth century; and, second, to explore how the unit's formation, work experiences, and postwar struggles were shaped by colonial politics—that is, how they were linked to perceptions that the NOFU was a unit from a fledgling colony that was a point of consternation for the UK government, which oversaw Newfoundland's public purse after the dominion agreed to suspend its own responsible government in 1934. My intention is to

11 In early 1945, the Registrar-General for Scotland, J. G. Kyd, reported that a total of 8,498 marriages between Scottish women and foreign nationals had occurred since the beginning of the war ("Canadian Beaus," 1945).

broaden the knowledge base about the unit in these regards and, in so doing, highlight some of the related intersections of war, work, migration, politics, and identity.

Along with Newfoundlanders and Labradorians, male foreign nationals from Canada, Australia, New Zealand, and British Honduras (now Belize), as well as the British Forestry Companies and the thousands of women in the Women's Timber Corps, also harvested and milled timber in the UK for the war effort. I provide some details of these other units as context and, where appropriate, as a basis for comparative analysis. As well, I juxtapose the experiences of the NOFU with its First World War precursor, the Newfoundland Forestry Companies. In so doing, I place the NOFU within a broader context of dominion war service within the forestry sector, to provide some comparative assessment of its contribution and experiences.

With an emphasis on documented research presented in a narrative form, my aim is to create an account that is both substantial and appealing, while also accessible to a variety of audiences that include members' descendants and family genealogists, as well as lay and academic historians in both Canada and the UK. Like *The Foresters' Scribe*, my earlier book on the Newfoundland Forestry Companies of the First World War, *From Loggers to Foresters* is a comprehensive account covering the NOFU's time in the UK as well as the pre- and postwar socio-cultural and political contexts of home. As is the case when writing for multiple audiences, specific details may be of more or less interest to some audiences than others.

Thematic Highlights

Given the archival material now available[12] about the NOFU, which reveals details of its formation, wartime work, and postwar efforts for recognition, a more thorough account is both needed and possible—one that can include previously unpublished and unknown details presented in the context in

12 Legislation guides the timeline for release of UK government documents into archival custody and availability for public viewing.

which they emerged. Little has been written about how colonial politics of the 1930s, including the introduction of governance by the NCG, as well as questions and debates about Newfoundland's future after the Second World War, shaped the NOFU experience in Scotland. This book addresses both these gaps and, in so doing, adds to accounts of the Second World War history of Newfoundland and Labrador.

Behind existing—and largely rose-coloured and partial—views of the NOFU and its accomplishments during the war lies evidence of tensions between the UK government and the NCG about the unit and its performance, as well as between the NOFU leadership and the branches of the UK government that oversaw forest production and wartime supply. This evidence demonstrates then-prevailing attitudes about Newfoundland colonials that were imbued with assumptions about masculinity and ethnicity widely shared in and by the UK press. It points to the tenuous grounds on which the work of the NOFU existed for the first years of the war, as it withstood a series of crises around its status, living and working conditions, and reception and reputation in the UK.

The hasty decision to strike the unit as a civilian outfit reverberated throughout the war and into the postwar period, threatening the unit's legacy and the self-efficacy and rights of the workers who formed it. While its large numbers reflected a historic wartime worker migration, the sheer size of this civilian unit created challenges in organization, supervision, and discipline that tested unit efficiency and diplomatic relations. Widespread UK press coverage about work styles, camp conditions, and after-work antics added to the unit's public relations problems. The situation reached a peak in 1942, a watershed year during which the unit's welfare, discipline, and member retention were improved, and prejudicial attitudes and behaviours of UK officials were named and challenged.

As the NOFU stabilized and advanced its reputation in the later years of the war, and as the question of the future of Newfoundland after NCG rule intensified, NOFU members considered their postwar well-being and the future of their homeland. Its member organization, the NOFA, was formed in 1944 for the benefit of members while overseas and upon their return home. It also had an explicit focus on the betterment of Newfoundland, seeking for its members to "take their full share in the work of making Newfoundland a country in which they and all other citizens may lead a good life" (NOFA, 1945, p. 68).[13]

As the end of the era of the NCG approached, members sought experiences—encouraged and supported by UK authorities through the British Council—that better prepared them to shape the future of their country. These activities arose not only from a long-overdue official concern for the well-being of the men, but even more from the increasing preoccupation of the Dominions Office and the UK government with smooth diplomatic relations in the context of the question of governance in Newfoundland after the war. As the spectre of a postwar economic decline in Newfoundland following its wartime boom threatened to present the UK with an additional fiscal burden along with its own extensive wartime debt, a desire to divest itself of its responsibility for the country grew, and the political as well as technical education of the men of the NOFU became of greater interest.

At the heart of the establishment of the NOFA was a belief in the potential collective impact at home of the knowledge exchange and adaptations that were an integral part of the men's work, life, and education in the UK. Much of what the men experienced overseas was new to them and many embraced opportunities to learn and advance their civic and political thinking, which, upon return, contributed to public discourse in a country in flux. Their leadership was needed—and tested—with their first fight: for recognition of their contributions through postwar benefits, a fight taken initially to the NCG and, after 1949, to the Government of Canada. It was a battle that would unfold over Newfoundland's first five decades in Confederation and offer the new province an enduring lesson in persistence in the pursuit of rights as Canadian citizens.

The total number of men who enrolled in the NOFU approximated almost 6 percent of the adult male population between the ages of 20 and 54

13 The organization is sometimes incorrectly referred to as the Newfoundland Overseas Forestry Association. The correct name—the Newfoundland Overseas Foresters' Association—appears in its constitution and monthly bulletins.

who were living in Newfoundland or Labrador in 1935.[14] Many of them worked with the unit for the duration of the war. Hundreds returned home with UK wives—the "war brides"—and new families. In their communities, this significant number of socially repositioned returning workers and new immigrant women with strong ties to their homes of origin contributed to debates on Newfoundland's future at the end of NCG rule, as well as to province-building and the evolution of citizenship after Confederation.

The NOFU story is noteworthy for what it reveals about the unit's place in the colonial politics of the time and the changing political landscape of home, and about workers' struggles for rights and recognition in these contexts and against the established hierarchies and meanings of war service within them. The fight for recognition of its contributions, so much a part of the NOFU's postwar story, was also present throughout the war, when the overseas men fought for a variety of improvements: timely payment of wages and access to leave; diets appropriate to the labour demands of their work; after-work amenities and opportunities to counteract isolation and boredom; and validation and respect for their skills and contributions. In the final years of its operations, with a stable nucleus of workers and concessions from the UK Ministry of Supply, the unit built a reputation as having disciplined workers and good citizens, thereby reshaping its legacy. Along with opportunities to pursue training and education, and to organize as skilled woods workers, seeking increased rights and responsibilities enhanced the men's potential for greater individual and collective status and political and social impact.

The achievement of the NOFU lay not only in its timber production quantity—which surpassed all other units—or the length of its service, which far exceeded that of any other wartime forestry unit. In the end, belief in the value of their contributions and knowledge, and a stalwart commitment

14 See Table A–3, "Population by Selected Age Group and Sex, Newfoundland and Labrador, for Selected Years 1935 to 1986," *Historical Statistics of Newfoundland and Labrador* (Newfoundland Statistics Agency, 1988).

to fight against the inequitable denial of rights and benefits, also emerge as key accomplishments. As this book demonstrates, the unit's experiences were framed by events that preceded the war and were rooted in and reveal much about the history and politics—as well as hierarchies of labour and location—of Newfoundland in the 1930s. But their efforts after the war also stand as a lesson in nascent citizenship in a new province, one in which wartime workers tested the meaning of their new country's democracy, and as a result contributed, along with others, to an expansion of the constituencies to whom benefits were owed and, in the process, to a redefinition of wartime service itself.

Book Overview

This book is divided into three primary sections.

The first—Pre-War Conditions and the Formation of the Newfoundland Overseas Forestry Unit—introduces the social and political context of Newfoundland in the 1930s during rule by Commission of Government, out of which the NOFU (and the Newfoundland response to the Second World War more generally) emerged. It focusses in particular on conditions in the woods industries, from which most of the men of the NOFU were recruited. It sets the emergence of the civilian NOFU beside that of its First World War military precursor, the Newfoundland Forestry Companies. In this section, the legislation, terms, recruitment, leadership, and transportation of the groups of contingents of the NOFU are also outlined.

The second section—Overseas Experiences: Work, Welfare, Organization, and Education—outlines the experience of the unit from its arrival in the UK to the repatriation of its last members in the summer of 1946. It provides analysis of the politics of their representation to and reception by the UK public, as well as some of the internal differences and points of tension among NOFU members themselves. It traces the early years of the unit and the crises of organization, productivity, leadership, well-being, legitimacy, and identity that plagued it. It also covers the less contentious final years, when the question of the future of Newfoundland was gaining political momentum. During this time, the NOFU secured better treatment by UK officials, developed a Home Guard unit, engaged in courses with the British Council, formed its own professional organization, and planned its return to a changing home country.

The third section—Repatriation and Reintegration: Postwar Struggles—covers the postwar period up to the beginning of the twenty-first century. This section explores how, during these decades, ex-NOFU members transitioned from wartime work into new lives at home. It reveals how the men who remained in the UK were eligible for postwar benefits and education, while those who returned to Newfoundland were shut out of postwar rehabilitation programs offered by the NCG. The efforts of the executive of the NOFA for recognition of the NOFU wartime contribution are traced. Many changes were afoot in these years, including within the woods industry, to which large numbers of NOFU members returned to work. The postwar era as a province of Canada also brought cultural, labour, and forestry changes, including further work-related migrations, in which the men of the NOFU participated.

Over the past decade or so, I have researched, written, and produced several projects about the woods workers of twentieth-century Newfoundland (Kelly, 2014, 2020, 2021b, 2022; Kelly & Forsyth, 2018, 2021). This body of research was inspired by a conviction that their place in the social history of Newfoundland and Labrador was inadequately documented, their labour contributions undervalued, and their expressive culture—in particular, occupational songs and stories—largely ignored. It was also impelled by a family connection to the industry through my father, who was a logger and camp foreman with the Anglo-Newfoundland Development Company (ANDCO). Woods work and the men who did it were the backdrop of my childhood years in Gambo, one of the towns at the centre of wood-cutting operations in rural central Newfoundland. These historic slights, therefore, felt somewhat personal, and researching woods workers and their cultural, economic, and wartime contributions as a form of remembrance and reparation was both disconcerting and rewarding. Theirs is a story of injustice and indifference, but it is also one of camaraderie and grit.

At the outbreak of war in 1939, my father, Andrew John ("Andrew J." or "A. J.") Kelly and his brother Thomas Leo ("Tom"), both unmarried, were working with the ANDCO. When not at camp, they lived at home with their ailing and widowed mother. My father, the youngest, remained at home with his mother during the war, while Tom joined the forestry unit, one of forty men from Gambo to do so. During his six years of work in Scotland, Thomas Leo Kelly (NOFU #1445)[15] was a camp foreman; he was also a Second Lieutenant in the 3rd Inverness (Newfoundland) Battalion. While I am familiar with many of the names of the men of Gambo who worked in the unit, Uncle Tom is my only personal connection to its story.

15 Assigned NOFU numbers are included at a member's first mention in this text. Appointed personnel, such as the Officer-in-Charge and Superintendents, were not assigned numbers.

SECTION ONE

PRE-WAR CONDITIONS AND THE FORMATION OF THE NEWFOUNDLAND OVERSEAS FORESTRY UNIT

Under the Cloak of Empire: Newfoundland in 1939

When the United Kingdom declared war on Germany on September 3, 1939, Newfoundland was not self-governing. Its responsible government—that is, governance in which an appointed governor and upper chamber are responsible to a legislature whose members are elected—had been suspended in 1934. In its place a Commission of Government was established, composed of Newfoundland's governor—the "Governor in Commission"—and six commissioners who were appointed by the UK government. Three of the commissioners were from the UK and three were from Newfoundland, with senior portfolios typically assigned to the former (Webb, 2008).[1] The NCG was overseen by the UK government through the Dominions Office, its Secretary of State, and other lower-ranking officials. Rule by the NCG resulted, in large part, from unmanageable national debt and near-economic collapse combined with political instability and corruption during the early years of the Great Depression (NRC *Report*, 1933). As a result of this status, when the UK declared war, it meant Newfoundland was also at war.

Newfoundland had demonstrated its commitment to the UK during the First World War, including through its Royal Naval Reserve and the raising of the Royal Newfoundland Regiment. That war had cost Newfoundland

1 Not all of the initially appointed commissioners served for the duration of the NCG. As with other forms of government, replacements and reassignments were common for a variety of reasons.

close to $35 million, approximately one-third of its overall debt burden in the 1930s; a large portion of the remainder of the debt came from the cost of building the Newfoundland Railway. Eventually the government was unable to make interest payments on this spiralling debt and maintain the everyday business of government, including care of its increasingly destitute people (Hiller & O'Brien, 2019; Noel, 1971; Webb, 2001). In an August 17, 1932, editorial, *The Western Star* described the political and business elites of Newfoundland as preoccupied with two options to resolve the country's debt—the sale of Labrador or Confederation with Canada.[2] It cautioned, "[Our people] must obtain relief from this intolerable burden, or perish" ("Sale of Labrador," 1932, p. 2).

Labrador, the mainland portion of the dominion that measures approximately 295,000 square kilometres (just under 114,000 square miles), had a population of approximately four thousand in the 1920s—many of whom were Indigenous Peoples, the Innu and Inuit—and it was seen by those in government as resource-rich and potentially lucrative.[3] During the 1920s, prior to a 1927 Privy Council decision that settled the Labrador Boundary Dispute between Canada and Newfoundland, three different Newfoundland Prime Ministers proposed its sale or transfer to resolve the debt problem: Richard Squires in 1922, William Warren in 1923, and Walter Munroe in 1924. In late 1931, in his second term as Prime Minister, Richard Squires once again offered Labrador to Canada for $110 million. By then, Canada was also wrestling with the financial and social effects of the Depression on its own people. Given these needs, financial support for Newfoundland was not feasible, Labrador was unaffordable, and any discussions about Confederation were untimely (Noel, 1971; NRC *Report*, 1933).

In 1932, as public agitation about government incompetence peaked in Newfoundland, a demonstration to protest government indifference and corruption under Squires became a flashpoint for desperate, destitute voters. It escalated into a riot during which demonstrators entered and damaged the

2 Andrew L. Barrett was editor of *The Western Star* from 1912 to 1941.

3 For perspectives on Labrador during this era, see NRC *Report* (1933).

Colonial Building, the seat of government. Squires resigned and Frederick Alderdice was elected Prime Minister. Alderdice belonged to the merchant elite, many of whom were ambivalent about responsible government. They believed uneducated Newfoundlanders and Labradorians were prone to political manipulation and were thus ill-suited for self-government (Webb, 2001).

Desperate to address the fiscal crisis, in exchange for last-ditch financial support from Canada and the UK, Alderdice agreed to a Royal Commission that would investigate the situation in the country and the options for action. With default on loan payments looming, and with both Canada and the UK fearing the impact that would have on themselves and other dominions, the UK government quickly appointed a group of three commissioners to complete the inquiry: Lord Amulree (William W. Mackenzie) of the UK, who would act as Chair, and Canadians Charles Magrath and Sir William Stavert. Their task was to examine the social and political circumstances in Newfoundland and to provide suggestions regarding its future (Cadigan, 2009; Hiller, 2002; Neary, 1973, 1988; Noel, 1971; Webb, 2001). A year later, in 1933, the *Royal Commission Report*, also known as the *Amulree Report*, delivered a devastating account of life in the dominion. Its recommendation: UK debt relief in exchange for suspension of self-government and, in its place, rule by a Commission of Government.

It was not only the careless additions to the financial debt to which the royal commissioners objected; in their report they highlighted political unscrupulousness, as well—"the greed, graft, and corruption which has left few classes of the community untouched by its insidious influences" (NRC *Report*, 1933, pp. 81–82), which included what it argued was the decline of moral character among those reliant on relief. In a sweeping indictment of its governance and political culture, the commissioners argued that Newfoundland needed "a rest from politics" (p. 195) and self-government. This rest was intended to offer political stability for "the country's wounded body politic" (G. Long, 1999, p. 78) as well as an opportunity for economic recovery. The report itself was not only a rationalization for this drastic measure; it was a charged version of the country and its people drawn from evidence presented to the Royal Commission, the impact of which reverberated for decades.

Rule by an elected commission as a response to Newfoundland's political troubles had been floated as an option a decade earlier by William Coaker, leader of the Fishermen's Protective Union, a political party as well as a workers' union. The Royal Commission recommendation, then, was neither novel nor, as the response to Coaker's idea had indicated, unsupported publicly (G. Long, 1999). Unlike Coaker's version, however, the one the legislature officially accepted in 1934 lacked any democratic elements whatsoever and, regardless of the degree of novelty or public support, was a rare political humiliation for a country whose situation was not unique among struggling constituencies during the Depression, including some provinces of Canada (Noel, 1971).[4] The inception of the unelected NCG reduced citizenship to "a low ebb" (Neary, 1993a, p. 85); without the right to vote, residents were without the central means of participation in politics.

During this rest from politics, which lasted from the mid-1930s until Confederation with Canada in 1949, the rebuilding of Newfoundland's economy was planned and directed by the NCG under the cloak of empire generally and, in particular, of the Dominions Office.[5] So, too, would be its contribution to Second World War service. Those who went overseas to serve during that time were drawn from the humiliated and downtrodden of Depression-era Newfoundland, who had been brought to their knees by destitution, illness, and government indifference. For two decades, many people had suffered poverty and powerlessness, while the losses of the First World War still hung over every community. When the Second World War was declared in 1939, the

4 In the debate on the second reading of the Newfoundland Bill in 1933 and in defence of Newfoundland, Clement Attlee, a Member of the British Parliament and Deputy Leader of the Labour Party, made this response to the possibility of its default on loan payments: "All the best countries default nowadays" (Attlee, 1933, col. 228). He provided several examples to support his point. Later, as Secretary of State for Dominion Affairs, Attlee became a key figure in the story of the NOFU and, as Prime Minister, in the story of Newfoundland's future after the Second World War.

5 The British Empire was a historic system of settlements, colonies, and territories claimed and dominated by Great Britain from the late fifteenth century onward, giving varying degrees of sovereignty and dependence to each. Following the First World War, it became a volunteer system of sovereign nations united within what is now called the Commonwealth.

country once again fell in line to support the UK, despite its own people lacking the franchise, enduring dire social circumstances, and bearing a still-fresh legacy of war remembrance. This time, however, its contribution would look different, but nonetheless be considered "stupendous" (Yarham, 1944, p. 15).

The First World War had enshrined war service as the ultimate expression of loyalty to empire; through service, valour might be rewarded with dignity (as partial compensation for loss)—at least for some.[6] But the experience of rule by the NCG had, by the time of the Second World War, weakened this sense of loyalty. Some felt that wartime service in the UK might be a way to repair it. As one NOFU member at the New Forest, England, told Admiral Sir Arthur Bromley of the Dominions Office in 1940 only a couple of months after the unit arrived in the UK: "You know we don't like you English as you took away our right of [self] government, but since we have been here, everybody has been so kind to us that we now have quite altered our minds" (Bromley, 1940, p. 1).[7] As time would tell, kindness was just one bridge needed to smooth relations with the NOFU and, by extension, between Newfoundland and the UK government.

Quick and Fulsome: Newfoundland's Response to the Outbreak of War

Despite having no military force of its own at war's onset—the Royal Newfoundland Regiment and the Royal Naval Reserve had been disbanded following the First World War—over 22,000 residents (from a population

---

6 *Daily Express* reporter Morley Richards visited Newfoundland in early 1939 and, on his return to London, wrote blistering accounts of the destitution and neglect that he found, prompting the NCG to write to the Dominions Office to address his points (see Neary, 2012). Two years later, Richards (1941) wrote that his report of a Newfoundland with "starkly tragic conditions" was now supplanted by a remarkable story of wartime sacrifice and "a record of selfless devotion to the Empire" unmatched by any other part of it (p. 1).

7 This dislike of the NCG was common among NOFU members, as demonstrated in letters about the future of Newfoundland published overseas in the *Newfoundland Foresters' Bulletin*. The NCG was also often the butt of jokes, which were intended to push back at its rule through humorous diminishment. See, for example, a joke retold by Patrick Ryan (NOFU #997) of Ferryland (Dooley, 1978).

of just over 320,000) participated in some form of Allied war service during the Second World War.[8] This figure includes those who enlisted in the home-based Newfoundland Militia (renamed the Newfoundland Regiment in 1943), the Royal Navy, the Royal Artillery, and the Royal Air Force. Newfoundlanders and Labradorians also served in the armed forces of Canada—navy, army, and air. As well, hundreds of women from the country served in the women's corollaries of these services. Most who served did so in a civilian capacity, including the 10,000 or so who were merchant mariners and the approximately 3,600 who were overseas woods workers (Higgins, 2007c).

Some of the initial decisions about how best to manage Newfoundland's war role and recruitment were tied to reluctance on the part of the UK government to incur expenses in addition to those of its existing caretaker costs for the struggling dominion (Nicholson, 1969). The NCG likewise wanted to avoid compromising any of the supports in place for the country's economic recovery. Its 1936 reconstruction plan, estimated to cost $800,000 annually, had won the approval of the UK government (Neary, 2011a), but a global recession shortly thereafter had curtailed its ambitions. Fiscal prudency was ever more important as the war began.

When a request for woods workers came from the Dominions Office in the fall of 1939, the NCG initially suggested it raise a military forestry unit, as had been done in the First World War. The shortage of training facilities was a hindrance and, as Commissioner of Natural Resources J. H. Gorvin would later suggest, the proper supervision to create such a military forestry unit might have been lacking at the time (Gorvin, 1940c). But the dire and pressing labour shortage in the UK was the decisive factor shaping the nature and formation of the unit. It would simply take too long to train a military corps (Eden, 1939; Gorvin, 1940c). A civilian unit could be raised more quickly, and it was a cheaper alternative to costly military preparation. It was the best way to address what Anthony Eden, Secretary of State for Dominion Affairs, called "an embarrassing shortage of labour" (1939, p. 1). Eden underscored the point by noting: "Coal production here, which is vital to war economy,

8 Of this number, more than 12,000 people served overseas, the largest component of which were the men who joined the NOFU (A. M. Fraser, 2010).

will be jeopardised unless production of home-grown pit props can be rapidly accelerated and any assistance which Newfoundland can offer in this matter would be very valuable" (1939, pp. 1–2). It was a rallying cry that would continue to ring throughout the first three years of the war.

A civilian unit was also a timely intervention in the massive unemployment and extreme poverty of the country. Throughout the war, the NCG would maintain a cold and stringent attitude toward the NOFU to ensure that continued to be the case. It would, for example, impose a government hard line on relief eligibility of men returning home after their first six-month contract ended and, at war's end, would deny ex-NOFU members access to re-establishment programs. Such policy exclusions revealed attitudes, including about the meaning of war service and who was eligible for reward through it, that would cause difficulties for the NOFU men both during the war and in the postwar period.

Along with fiscal concerns, the urgency of providing a means for war support further shaped the nature of the unit. At the outbreak of war, the NCG was quick to indicate its eagerness to supply combat power, and Newfoundlanders and Labradorians were eager to participate (as they had been during the First World War). On September 7, 1939, four days after war was declared, the Governor of Newfoundland, Sir Humphrey T. Walwyn, indicated as much in a telegram to Eden, who responded that Newfoundland volunteers would best be trained in the UK for entry into its forces there (Eden, 1939). Slow progress in authorizing recruitment, however, frustrated the enthusiasm to serve. In a telegram to Eden on October 3, 1939, Governor Walwyn noted: "Public opinion is being worked up because of apparent inactivity of Government in relation to our war effort. It is therefore of great importance that we should be enabled to make some pronouncement at the earliest possible moment" (Walwyn, 1939).

The formation of the NOFU a few weeks later and the ambitious call for men to join it offered a release for the pent-up urge to serve in any capacity—in this case, to get overseas as waged workers and, at end of the contract if so desired, pursue opportunities in the combat forces of the UK. As it happened, recruitment for the Royal Navy began in late October; at first, just for 625 special services sailors, but shortly thereafter for 1,000 volunteers for general services.

Recruitment for the Home Defence Force (later renamed the Newfoundland Militia) also began in late October; initially it attracted few volunteers because of limited training facilities and no barracks. Recruitment for the Merchant Navy did not begin until the spring of 1940 (Nicholson, 1969).

Damnable Conditions: Logging in Newfoundland in the 1930s

Conditions in Newfoundland's woods industries in the late 1930s were a factor in the successful recruitment of workers into the NOFU. In 1936, approximately six thousand loggers were employed in the forestry sector (Walwyn, 1936, as cited in Neary, 1985), most with the two large, English-owned pulp and paper companies that operated in Newfoundland: the International Pulp and Paper Company at Corner Brook, purchased by Bowater-Lloyd in 1938, and the Anglo-Newfoundland Development Company, which had opened at Grand Falls in 1909.

The 1930s was a decade of much woods labour unrest, with job actions occurring across the Island and in Labrador.[9] Right from the advent of the

A logger stands atop a woodpile at Terra Nova, Newfoundland, ca. 1938. (Courtesy of The Rooms Provincial Archives Division, VA 53A–39.1)

9 For an account of this decade of unrest, see Sutherland (1992). Sutherland tallied that there were fifty labour disputes in the woods industry between 1929 and 1939.

pulp and paper industry in Newfoundland in the early twentieth century, loggers had not been well paid. The conditions under which they lived and worked were also deplorable: long hours of physically demanding and dangerous work; uneven access to good cutting "chances" (sites), which affected wages that were based on cordage; poor diet; crowded, uncomfortable, and lice-infested camp accommodations; and a system of work organization that left them vulnerable to the interests and bidding of the companies that profited from their labour.

The Logging Act of 1915, championed by William Coaker, had set some industry standards for living conditions and food and, in 1931, a minimum wage clause was added. The legislation was, however, largely ignored by employers who, in the face of declining exports in economically difficult times, protected their profit lines, unchecked, on the backs of workers (Sutherland, 1992). Many experienced loggers abandoned the industry as the decade wore on in response to declining prices for harvested cords of wood and poor working conditions. To replace them, both companies hired workers who had little or no experience; these men, too, often quit when they realized their pay did not cover their work costs and left nothing from their earnings for their families. Some of them were even arrested and imprisoned when they were unable to pay the fare for their trip home from camp (No Oil in Our Lamps, 1934). According to the *Amulree Report*, "well-nigh desperate" were the woods workers of the International Pulp and Paper Company, while those who worked for the ANDCO were deemed able to cover their living expenses but with a small margin (NRC *Report*, 1933, p. 150).

Against this backdrop of difficulty and unrest, in 1934 the NCG ordered an investigation into the conditions of work and life at woods camps, the terms of employment, and any other matters related to woods workers.[10] The inquiry was led by F. Gordon Bradley, a lawyer, magistrate, and politician. He travelled widely along the railway and coastlines, meeting with an array of people who testified about the conditions of woods work in the country. Bradley recorded with clarity and precision the shocking evidence he

10 For a discussion of these conditions, see Sutherland (1991).

collected, concluding that "the woodsman has been reduced to a standard below even tolerable existence" (Bradley, 1934, as cited in Neary, 1985).

Those who worked in the mills and lived in the company towns, however, suffered no such indignities. The paper companies built, through their policies, a cushion of paternalism against dissent; they purposefully divided the industry and workers by creating a nascent class difference based on work type and location. They denied woods workers access, for example, to company town residency. As logger John Kitchen observed in his memoir, a class hierarchy emerged among workers in the pulp and paper industry: at the top were millworkers, followed by personnel of the various depot towns; and at the bottom were the woods workers (Kitchen, 2005).

It was not only social class that divided the people working in the logging and lumbering industries. Indigenous Peoples on the Island and in Labrador, whose lands were ravaged by unchecked and careless company capitalism, were treated as a subclass on the lowest rung of workers, and they faced unrelenting additional harassment and discrimination because of their culture and identity (Hanrahan, 2013). Ignatius Paul (2000) recounted how the ANDCO relied on Mi'kmaw knowledge of the land and waterways to transport men and drive pulpwood at Badger, yet the Mi'kmaq men themselves were not allowed inside the homes of those for whom they worked. As Norman Gabriel (NOFU #0036) of Stephenville described it: "Because we were Indians, I was treated like the plague" (Gabriel, 1999, para. 2).

The NCG refused to publish the *Bradley Report*, fearing it would incite even further unrest among woods workers and antagonize the pulp and paper companies, whose interests it preferred to protect. It did, however, use some of Bradley's findings to negotiate with the companies for better conditions, including a commitment that they would pay loggers a minimum wage of $25 per month (Bradley, 1934, as cited in Neary, 1985).

It was difficult for the loggers, however, to challenge the companies about the conditions outlined by Bradley. Not only were they among the poorest of the poor, they were also poorly organized—in part because of how the work itself was organized: in isolated camps in the Island's interior, where the men were disconnected from one another and success was based on individual performance measured by quantity cut. Nevertheless, the camaraderie that was a hallmark of the occupation was built within these camps, in bunkhouse settings where men shared their stories of work and often composed songs and poems to document their experiences. The existing texts of these compositions demonstrate the critical acumen of workers, as well as how they were often frustrated in their efforts to advance their cause.[11]

In 1935, the first exclusive organization of woods workers, the Newfoundland Lumbermen's Association, was founded by Joseph J. Thompson. Previously, loggers had been represented along with fishers by the Fishermen's Protective Union, which had made some inroads on their behalf but suffered waning influence by the 1930s (Sutherland, 1992). By the end of the decade—and on the advent of war in Europe—there were several other players in the organizational matrix of woods workers in the country. Two were splinters groups of the Newfoundland Lumbermen's Association—the Newfoundland Labourers Union and the Workers Central Protective Union—that had emerged following dissatisfaction with progress made by Thompson. In 1940, the NCG formed the Woods Labour Board, which brought together pulp and paper company representatives and union leaders to facilitate and expedite resolution of the issues brought forward there. The Woods Labour Board offered much needed stability in the sector, especially during wartime; while it ultimately worked in favour of the companies and the NCG, it also placated loggers who wanted their concerns to be heard by both parties (Gushue, 1956; Sutherland, 1995).

The attitude of the government regarding forest resources and workers within the forestry sector was muted. In a lengthy despatch to Secretary of State for Dominion Affairs Malcolm MacDonald at the end of 1936, the NCG presented its reconstruction plan for the country to the Dominions Office. Governor Walwyn outlined the plan as a series of responses to questions from MacDonald, who was gathering information on economic issues for use in policy formation. Regarding forest resources, Walwyn wrote about the need for greater knowledge (in particular about Labrador forests), closer

11 For an account of this cultural legacy, see Kelly (2014) and Kelly & Forsyth (2018, 2021).

administration, and fuller use. He added that the forests of Newfoundland and Labrador "afford hard and ill-paid labour to a section of the population," but it was unlikely they would contribute significantly to economic reconstruction in the near future (Walwyn, 1936, as cited in Neary, 2011a, p. 255). The possibility of a third mill and the employment boost it would offer, although still alive when Walwyn wrote, was foreclosed in 1938 when Bowater negotiated a deal to acquire additional timber rights and expand capacity of their Corner Brook mill rather than build the previously proposed new mill on the Gander River.[12]

When recruitment for the NOFU began in late 1939, the terms of employment overseas no doubt struck many of the loggers who had lived through the deprivation, chaos, and uncertainty of the 1930s as offering stability and predictability—and a way to manage some form of income to help their families while also supporting the war effort. Praising the terms of the initial contract, the editor of *The Evening Telegram* wrote that the contract "should prove highly attractive and will remove from many a home the dread of a winter of inactivity for the provider of the household" ("Nfld. Forestry Unit," 1939, p. 6).[13] Under its terms, workers were required to allot 50 percent of their overseas income to their families as a guarantee that they would be provided for during the separation period.

The model—emigration in order to get work and allotments from wages sent home—had a familiar ring. It reflected a historic pattern: people left the country during hard times and sent money home to their families while they were away. The Great Depression had drawn many such emigrants back, however, and had discouraged others from leaving, given that employment opportunities were just as limited elsewhere as they were at home. Their return had increased the burden on the public purse (NRC *Report*, 1933). The establishment of the NOFU was in some ways a formalization of this historic pattern, but this time it was intended to ease the strain on the NCG as well as provide a supply of cash to destitute families.

---

12 For an insightful discussion of these negotiations, see Neary (1988).

13 Charles Jeffrey was editor of *The Evening Telegram* from 1923 to 1959.

The Overseas Hewers of Wood: A First World War Precursor

The NOFU assignment in the UK—to cut wood to support the Allied war effort—had a precursor in the Newfoundland Forestry Companies of the First World War. The circumstances that resulted in the formation of such units in each war were similar—a UK shortage of home-grown timber and skilled woods labour against a backdrop of overreliance on imports threatened by sea warfare. Along with forestry units from Canada, Portugal, Finland, and the United States, and local workers including the Women's Forestry Service, the Newfoundland Forestry Companies helped meet timber needs during the Great War by creating a local supply and felling trees on private estates in the UK.

The Newfoundland Forestry Companies military unit was formed in spring of 1917. Comprised of approximately five hundred soldiers, the men wore the uniform of the Newfoundland Regiment of which their unit was a part. To be eligible for enlistment, recruits had to be ineligible for other service units. There was no upper age limit and, when the company was at strength, its soldiers were aged 14 to 70 years. About two-thirds of the men who volunteered came from rural communities outside St. John's.

Members of the Newfoundland Forestry Companies, Craigvinean, Dunkeld, Scotland, 1917. (Courtesy of The Rooms Provincial Archives Division, VA 55–5.1)

The work of the Newfoundland Forestry Companies entailed a combination of military drills and wood harvesting and milling. Their task was to fell private forests on the Scottish estates of the Duke of Atholl at Craigvinean Hill near Dunkeld, and those of the Marquis of Breadalbane at Drummond Hill near Kenmore, both in Perthshire in central Scotland. The unit was divided into three companies that served together in these small locales. (The NOFU, a much larger unit, would be divided into camps and districts, dispersed initially in England and Scotland, and then later concentrated in the Scottish Highlands.)

The Newfoundland Forestry Companies operated for just under two years and was disbanded in August 1919; most soldiers had been repatriated during the preceding winter. By all official accounts, their service was an invaluable contribution to the success of the war effort. Those who served were eligible for veteran pensions and allowances, although their experiences in securing fair allotments at the end of the war and the disparaging attitudes they faced about the meaning of their service portended somewhat how the men of the NOFU would be viewed and treated on their return, a quarter of a century later.[14]

While the overseas context for the formation of the wartime forestry units of Newfoundland was similar, the home context of each was dramatically different. The men who volunteered for the Newfoundland Forestry Companies during the Great War left a country that was enjoying wartime prosperity (albeit short-lived) under an increasingly divided party system of responsible government (Hiller & O'Brien, 2019). Recruitment began mid-war in a country that had grown tired and suspicious of the conflict and its human and financial costs. Enlistments were slow, with the unit reaching only half of the targeted size of one thousand soldiers (Kelly, 2020). Those who signed contracts for the NOFU, on the other hand, left a struggling country emerging from the Great Depression and ruled by the unpopular and appointed NCG. They were eager to sign for overseas work.

This combination of political and social unrest and destitution in Newfoundland and the dire trained-labour shortage in the UK was the backdrop for

14 For a detailed account of the history of the Newfoundland Forestry Companies, including issues of recognition and reward, see *The Foresters' Scribe* (Kelly, 2020).

war recruitment in late 1939. In the final analysis, the largest group of Newfoundlanders to serve overseas during the Second World War—and the first to arrive there—were neither sailors nor soldiers. They were sawyers and loggers, the vast majority—close to 90 percent—of whom left rural communities at home to become the civilian workers of the Newfoundland Overseas Forestry Unit.

A Civilian Workforce: The Newfoundland Overseas Forestry Unit

Following initial limited recruitment for the Royal Navy in October 1939, the NCG made an urgent public request for woods workers on November 9 (Eden, 1939). A request similar to the one sent to Newfoundland by the Dominions Office was also sent to other dominions that did not respond with the same swiftness as Newfoundland. For example, it would be summer 1940 before the largest unit from the dominions, the Canadian Forestry Corps, was established and mobilized for service, and winter 1941 before most of its companies began arriving in the UK (Wonders, 1987, 1991). But Newfoundland was motivated to respond quickly to the urgency expressed by the Dominions Office, and its civilian unit required less preparation for overseas work than the Canadian Forestry Corps, a large military outfit. Following consultation with woods managers at both the ANDCO and Bowater regarding the potential impact of recruitment on their workforce, plans for establishing the unit began.

An Act Respecting the Newfoundland Overseas Forestry Unit was passed on November 18, 1939. It confirmed the official name of the unit—the Newfoundland Forestry Unit, different from what appeared in the legislation's title. Unlike the initialism of its official name (NFU), using NOFU avoided any confusion between the forestry unit and the National Farmers Union, a concern Governor Walwyn raised when a badge for the NOFU was being designed (Walwyn, 1940). The NOFU was funded by the UK government, but the Act clearly legislated that the unit was overseen by the NCG through its Department of Natural Resources and led by a Chief Overseas Forestry Officer appointed by the NCG. Once in the UK, the Officer-in-Charge liaised with the Chair of the UK's Forestry Commission and, later, its

Ministry of Supply regarding matters of the unit's operations in the UK (*An Act*, 1939).

On the evening the Act was passed, November 18, 1939, Newfoundland's Commissioner of Natural Resources, J. H. Gorvin, addressed the country and called for two thousand workers to cut timber for pit props in England and Scotland as part of the newly minted NOFU. In his radio broadcast, Gorvin first spoke to appease the eager volunteers for whom recruitment calls seemed too-long coming, noting that the UK fighting forces were "on a much stronger footing at the outbreak of this war than in 1914, so that the call for recruits has not been so insistent" (Gorvin, 1939, p. 3). The need for timber cutters, however, was urgent.

In a carefully crafted argument to impress upon potential workers the significance of civilian wartime forestry, Gorvin underscored lessons from the previous war, in particular, "the excellent reputation established in Great Britain by Newfoundland woodsmen during the last war and held ever since" (1939, p. 6). Fearing, or perhaps anticipating, a dissatisfaction with the civilian status of the unit, he pointed to the importance of keeping skilled men in their work if they could be spared from the front line, and emphasized what would become a retention-directed refrain to men in the NOFU camps across the UK during the war: "The greatest contribution each man can make lies in doing the work for which he is best fitted" (1939, p. 3).

The First Contract[15]

Following Gorvin's speech, the first call for workers appeared in local newspapers, and recruit teams moved into designated areas soon after. The call targeted men aged 18 to 50 who met the physical health standard for work in Newfoundland woods camps. It outlined the terms of engagement: a six-month contract, with the option to renew or repatriate upon its expiration. Free return transport between Newfoundland and the UK was guaranteed each man. Basic pay was set at the minimum wage used by the ANDCO and Bowater: $2 per day in a six-day work week, with higher wages for foremen and sub-foremen.

15 Contracts and terms were widely publicized in local newspapers. For copies of the first and subsequent engagement contracts and renewals, see T. V. Curran (1987).

A required allotment of 50 percent of pay would be designated for next of kin in Newfoundland. Additionally, meals, accommodations, and medical care would be provided. Workers would purchase their own clothing, but equipment, such as axes and bucksaws, would be provided by the woods companies.

In signing, each man agreed to work industriously, comply with orders, and behave in a law-abiding and orderly manner. Persistent misbehaviour would be punishable by discharge and the cost of return transport would be at the discharged member's expense. Lost time due to sickness would be paid at half rate, while the Employers' Liability Act and Workmen's Compensation Act would guide medical care and compensation. Workers deemed medically unfit to continue overseas work would be transported to Newfoundland free of charge (Gorvin, 1939).

A newspaper editor with *The Evening Telegram* in St. John's praised the contract, writing that it offered "assurance of steady work for a large number on a remuneration basis that should prove highly attractive" ("Nfld. Forestry Unit," 1939, p. 6). This optimism was shared by others. In a year-end editorial, the editor of *The Bay Roberts Guardian* noted how much more promising the country's economic outlook was at year's end than at its beginning. He cited the bump in employment numbers created by NOFU workers and those who replaced them in jobs vacated at home, as well as the economic benefits of the allotments sent to their home communities ("Review and Prospect," 1939).[16] *The Evening Telegram* editor also noted how the age range of eligibility for work opened participation to those "who could not hope to serve their King and Country in the theatre of active warfare" ("Nfld. Forestry Unit," 1939, p. 6).

The first NOFU contract also stipulated that, at expiration, workers had the option to enlist in an armed service or another unit, provided reasonable notice was given and permission granted by the Officer-in-Charge. Considering that few other options were available for armed service when the NOFU was created, and that the extent of the need in the UK was unclear and the duration of the war unknown, this clause was not surprising—it positioned joining the NOFU as a way to start contributing to the war effort and possibly

16 Charles R. Russell was editor of *The Bay Roberts Guardian* from 1939 to 1944.

as a first step, a point Turner later acknowledged was the case for many of his men (Turner, 1940c). This positioning would change over time. The Secretary of State for Dominion Affairs, Viscount Caldecote,[17] would later indicate to Governor Walwyn his gratitude if recruiters for other services could be advised "to exercise special caution" in accepting loggers, a move that reflected both the persistent demand for woods labourers and emerging retention issues in the NOFU (Viscount Caldecote, 1940a).

Caldecote's advisory, combined with new restrictions on transfers to combat forces in the contracts offered to woods workers in 1940 and 1941, along with protracted pressure by UK and NCG officials on men to remain in the unit, effectively shut the gate on the NOFU being a route overseas toward enlistment in other services. It also streamlined loggers into the unit that the authorities felt best fitted their skills and Allied needs. The renewal contract for subsequent groups offered men, in addition to a six-month extension, longer work-term options: one year or "for the duration." It also clarified expectations for the period during which time-expired workers awaited transportation home.

In a retrospective analysis, newspaper columnist and First World War veteran Frederick Matthews called the NOFU contract "inhospitable" (1966, p. 16) and glaringly devoid of any postwar provisions of care for the men who signed it. Unless they had transferred from the NOFU to an armed service, in which case military pensions were assured, the first and subsequent contracts promised the men no postwar benefits. The contract was prepared in haste, while the men were being recruited and gathering for the rush overseas. There was little time to consider such matters in the face of the urge and surge to serve, but there was also no reason to expect such consideration from the NCG. Had the NOFU workers foreseen their postwar treatment, many more may well have taken the opportunity to transfer out when they could have, a point that would not be lost on those who would later make the case for service recognition to Canada's Department of Veterans Affairs (VAC) in the postwar period (Canada, Parliament, House of Commons, Standing Committee on Veterans Affairs, 1960).

---

17 The 1st Viscount Caldecote was Thomas W. H. Inskip.

Additional Calls and Contracts

In 1940, when the first NOFU contracts expired, a wave of requests to transfer to armed services was expected, given the enthusiasm to serve and the by-then increased options for doing so in a uniformed—and combat—role. From the perspective of the UK Forestry Commission, which desperately wished to keep workers occupied in essential woods work, this wave was more like a devastating tsunami. In total, when their first contract ended, of the just over 2,150 men who had signed (minus those who had been returned to NF for medical or disciplinary reasons), 267 transferred from the NOFU to the Royal Navy, 219 to the Royal Air Force, and 47 to the Royal Artillery (A. M. Fraser, 2010). Hundreds more returned to Newfoundland, where the wartime economy was now offering more and better employment opportunities. The need to replace these workers was both dire and complicated.

The economic context facing recruiters in Newfoundland had vastly changed since 1939. First, demand for local labour had surged thanks to the construction of airports built by the Allies who recognized Newfoundland's essential location for the defence of North America (High, 2003; Neary, 1988). As well, the pulp and paper industry, which had suffered a downturn before the war, had rebounded. In talks with the pulp and paper companies, the Department of Natural Resources was warned of the potential impact on labour supply of extensive recruitment. Both parties agreed that an additional one thousand workers could be spared, however, and a second call was initiated in spring of 1940 (C. Fraser, 1940). Recruitment of another 450 men followed in summer of 1941. During this third call, the drive for workers was extended to previously unreached locales, including the Island's Northern Peninsula and its south coast ("Arrangements," 1941).[18] Ex-NOFU members who had returned to Newfoundland were not eligible to re-enlist ("Men Wanted for the Newfoundland Overseas Forestry Unit," 1941).

The contract offered to later groups of NOFU recruits included clauses intended to stem attrition. The Forestry Commission, who claimed it had not

---

18 Only a few men from Labrador joined the NOFU. For information about those who did, see "Labrador's Veterans" (2020), an issue of *Them Days* magazine.

been consulted on the original six-month term and considered it too short, argued that "what we really require is for the men to remain with us for the duration" (Robinson, 1940b, p. 2). Workers who responded to subsequent recruitment calls were thus required to sign up for the duration of the war, and transfers to the armed forces or another unit were not generally permitted. To address problems of disorderly conduct and inadequate work, which had emerged among the first group, the later contract included stronger language about fines and dismissal. It gave the Officer-in-Charge an option to retain up to $25 of wages to cover repatriation transportation in the case of dismissals. This contract also outlined income tax and worker compensation requirements and included a stipulation for a clothing allowance. While the contract stated there was no pay for work lost due to weather—which the original group received—this clause was later revoked to create parity among all contract groups (Dunn, 1941a).

The men recruited in the second and third calls assuaged the labour shortage created by the hundreds of transfers and non-renewals at the end of the initial contract. But they did not fulfill the need for woods workers completely. At that point in the war, prior to the Battle of Dunkirk, the War Office was planning twenty-eight forestry companies to be attached to the Royal Engineers and posted on the Continent. Viscount Caldecote pointed out that these companies offered "the best opportunity for making military use of special skilled Newfoundland loggers" (Viscount Caldecote, 1940a, p. 1), and they would also meet the men's wish to contribute in uniformed service. The issue of uniforms had emerged with great fervour after the NOFU's arrival in the UK, when the limits of their status as civilians became glaringly obvious. This plan, however, was scuttled as the German forces advanced through Europe, France fell, and the Battle of the Atlantic grew more intense.

By early fall 1941, as both the urgent need for a reliable timber supply and the labour shortage escalated, Viscount Cranborne, who had replaced Caldecote as Secretary of State for Dominion Affairs, wrote again to Governor Walwyn to raise the possibility of recruiting a second unit of two thousand workers.[19] This unit, the Ministry of Supply proposed, would be organized along the same lines

19 Viscount Cranborne was Robert Arthur James Gascoyne-Cecil, 5th Marquess of Salisbury.

as the NOFU and have its own Officer-in-Charge, the existing unit being "as much as Captain Turner can effectively manage" (Viscount Cranborne, 1941b, p. 1), perhaps a suggestion of some emerging lack of confidence in his leadership.

By this time, camp conditions for the NOFU workers were well-known at home. As Peter Dunn, who replaced Gorvin as Commissioner of Natural Resources, wrote: "Reports here of inadequacy of certain food supplies in present logging camps might be [a] deterring factor in new recruiting" (Walwyn, 1941a, p. 1). Food-related issues—meagre war rations in the context of demanding physical work and a lack of canteens for after-hours food access—were not the only negative factors.[20] The questionable welfare conditions of the NOFU in the UK were also becoming known and deterring interest. Citing increased labour demands at home and NOFU members' reluctance to pay UK income tax (Walwyn, 1941b), the NCG was unable to find workers to fill a 1941 request by the Dominions Office for employment of four hundred sawyers who would not be members of the NOFU ("Men Wanted," 1941). Given these developments, raising an entire second unit was unlikely to succeed.

Additionally, the decision to create the NOFU as a civilian unit had created internal issues for maintaining morale, discipline, and productivity—so much so that the unit's reconstitution as a semi-military unit was considered (Gorvin, 1940c). Later in the NOFU's overseas work term, amidst a crisis of relations between the Minister of Supply and the unit, Minister Andrew Duncan commented to then Secretary of State for Dominion Affairs Clement Attlee that "the status of the Unit was unsatisfactory in that a large body of men was sent over to the United Kingdom without the cohesion or standing of a military unit, and the discipline and privileges which this would have carried with it" (Duncan, 1942, p. 2).

NOFU Leadership: Vision and Experience

Shortly after the Act to form the NOFU was passed and the initial contract terms finalized, the unit's leadership was announced. In the First World War,

20 Canteens and recreational huts were not common in NOFU camps until the summer of 1942. Canteens (or "British Restaurants") were places where workers could get light snacks after work hours.

the method of choosing leadership for the Newfoundland Forestry Companies had been contentious.[21] Some local papers ran letters and editorials contesting appointments of some commissioned officers, charging the authorities with cronyism and overlooking local men of skill and experience in the woods industry in order to favour well-connected men "from away" ("Alas! The Poor Native," 1917). There were no such public debates about the initial leadership appointments to the NOFU. All appointees had extensive knowledge of forestry and woods operations and held leadership positions within the Department of Natural Resources or the pulp and paper companies.

The NCG appointed as Chief Overseas Forestry Officer, or Officer-in-Charge, John ("Jack") Turner, who was at the time the face of forestry leadership in the country.[22] Turner was from St. John's and had worked with both the Anglo-Newfoundland Development Company at Grand Falls and the Reid Newfoundland Railway before moving to British Columbia in 1911. He served in the Canadian Expeditionary Force during the First World War, reaching the rank of Captain, and was awarded the Military Cross.[23] Following the war, Turner returned to Newfoundland and worked as a forest surveyor and timber cruiser with the ANDCo. He also worked with the Anglo-Canadian Pulp and Paper Company in Quebec

Captain Jack Turner in the uniform of the Canadian Expeditionary Force, 1918.
(Photo from "Zero Minus One" score sheet; lyrics by Jack Turner, music by William Fisher)

21 For an account of this issue, see Kelly (2020).

22 In his capacity as Officer-in-Charge of the NOFU, Turner did not use a designation in correspondence, but he was widely referred to as "Captain Turner" in Dominions Office correspondence and elsewhere. He was assigned the rank of Lieutenant-Colonel upon formation of the 3rd Inverness (Newfoundland) Battalion of the Home Guard, a NOFU-constituted unit which he led.

23 Once back in Newfoundland after the Great War, Turner penned poems about his wartime experiences. See Turner (1917).

(Baker & Miller Pitt, 1996). In 1934, Turner was appointed the Forestry Officer with the NCG's Department of Natural Resources. In this position, he developed innovative, experimental logging, sawmilling, and tree nursery operations on the Avalon Peninsula's Salmonier Line, at Back River and Deer Park, that employed hundreds of woods workers and modelled improved working conditions during the Great Depression (Powers, 2000). These efforts were intended to address the concerns of woods workers that had been outlined in the *Bradley Report* (Bradley, 1934, as cited in Neary, 1985) and more general concerns about sustainable forestry practices, some of which were identified in the *Amulree Report* (NRC *Report*, 1933).

The value of Turner's work at home was not lost on some, including the editor of *The Evening Telegram*, who questioned the repercussions for the Newfoundland woods industry of Turner's overseas appointment, and the wisdom of dispensing with his services for an extended period given "the importance attached to the Forestry Department generally, and in view as well of the increased need of supervision of operations" ("Nfld. Forestry Unit," 1939, p. 6). Turner was not the only drain on woods leadership in the country exacted by the needs of the unit.

Along with Turner, Governor Walwyn appointed initial superintendents Edgar Baird, Joseph Curran, Bruce Fraser, and Alleyne Hubbard ("Appointments," 1939; "Public Notice," 1939). Edgar Baird of Campbellton had been Chief Woods Ranger for Newfoundland, as well as a logging superintendent at Glenwood for the International Pulp and Paper Company (later Bowater) and a district manager for the Island Timber Company (Baird, 1981). Joseph Curran of Gambo, along with his younger brother Thomas Curran, worked with Turner at the Deer Park and Back River logging operations (Powers, 2000). At the time of his appointment, Joseph was Superintendent of the Game Reservation at Salmonier Line. Both Currans also had experience in family timber operations at Gambo. Born in Nova Scotia, Bruce Fraser (brother of Claude Fraser, who was the Secretary of Natural Resources in the NCG), was a forestry engineer with Bowater-Lloyd, which had purchased the International Pulp and Paper Company mill at Corner Brook in 1938 ("News About People," 1947). Alleyne ("Al") Hubbard of Burton, New Brunswick, a professional forester, was a graduate of the University of New Brunswick with a Bachelor of Science (Forestry). In 1937, he had moved to Newfoundland for work with the International Pulp and Paper Company at Corner Brook (Jarratt, 2016).

These initial appointments saw the NOFU through its transition from recruitment to overseas operations. Once in the UK, the challenges of leading and supervising a large unit of civilian workers and at the same time satisfying the standards of the Forestry Commission—and after 1941 the Ministry of Supply, the UK government departments under which the unit fell—were quickly apparent. With subsequent reorganization for increased supervision of workers (at the insistence of the Forestry Commission) and in response to personnel changes in the first years, additional superintendents were appointed. They included Charles Cahill of St. John's, Brendan Davis of Brigus, Clarence Lingard of Bishops Falls, James G. Martin of St. John's, and Peter Petipas of Whitbourne. The Forestry Commission Chair, Sir Roy Robinson, considered supervisory appointments from outside the unit and the country, but no such appointments were made (Robinson, 1940c).

Those in leadership positions were required to supervise; they also were committed to fair treatment of the men. Their tolerance for the Forestry Commission's blatant criticisms and disrespect likely challenged their dedication to the unit; both the Officer-in-Charge and the initial superintendents, as well as some foremen and sub-foremen, came under regular fire from Robinson. For his part, Robinson assumed the NCG supported him in taking any actions necessary to improve the unit's supervision. To curry its favour, he wrote to Gorvin: "The Forestry Commission fully appreciates the anxiety of the Commission of Government that the Newfoundland Forestry Unit's achievements shall be in line with other parts of Newfoundland's splendid war effort" (Robinson, 1940e, p. 2).

"The Country Cannot Afford Shirkers": Recruitment and Relief

The response to the initial call for wartime woods workers was enthusiastic. Some men even expressed interest by telegraph on November 18 while Commissioner Gorvin was addressing the country on radio (Gorvin, 1939). Personnel in the Forestry division of the Department of Natural Resources led recruitment; given the imminent onset of winter and the increased travel difficulties it would bring, their efforts focussed on the more accessible parts of the country—the east, west, and central portions of the Island. They were extended to more remote areas when subsequent calls went out (C. Fraser, 1940). For the locations to which recruiters were unable to travel, the Newfoundland Rangers provided assistance. A quasi-police force formed by the NCG in 1935, the Rangers provided administrative government services in remote parts of the country, including Labrador, where the Newfoundland Constabulary—the country's actual police force—did not have an established presence (Higgins, 2015).

In the First World War, recruitment efforts for the Newfoundland Forestry Companies had been much less robust and many of the recruits far less

A Newfoundland Ranger on a recruitment drive at a Newfoundland woods camp, 1939.
(Courtesy of The Rooms Provincial Archives Division, A 7–166)

experienced. Some who enlisted were Regiment soldiers who were no longer considered fit for combat roles. Many were men unfamiliar with woods work; they received basic training in logging and lumbering techniques, as well as military training, prior to travelling overseas. Their lack of expertise drew criticism; one experienced lumberer wrote that "it shows a gross lack of knowledge on the part of those in authority" to enlist office workers and store clerks for the work of logging and lumbering (Simmonds, 1917). The unit's expertise was also questioned while it was in the UK.[24]

When the Second World War began, the context for recruitment in the country had changed. With two large pulp and paper mills now operating and thousands of men involved in and knowledgeable about woods work, the recruit pool was much larger and more experienced. Volunteers who demonstrated the required knowledge about logging and passed a medical exam then signed a contract accepting the conditions outlined in the call, swore an oath of allegiance to King George VI, and awaited transport overseas. While measures to ensure skill levels appeared to be in place, their rigour and the workers' readiness were quickly called into question in the UK (Eden, 1940; Turner, 1940b). Sir Roy Robinson of the Forestry Commission commented that many in the first group sent overseas in late 1939 through winter 1940 "were unsuited by character, vocation or outlook for the venture" (Robinson, 1940e, p. 2).

To the initial general enthusiasm to enlist, the NCG added its own pressure. In the spring of 1939, the government was supporting well over 84,000 able-bodied dependents, a number that had increased with the economic downturn of 1937 that had strongly affected the pulp and paper industry (Neary, 1988, 2012). Seeing an opportunity to decrease the number of people on relief—or "the dole" as it was commonly called in Newfoundland (Story et al., 2006a)—through wartime work in the UK, the Department of Public Health and Welfare sent a notice to all Relief Officials on December 5, 1939, outlining the expectation that physically fit men between the ages of 18 and 50, married or single, must be denied relief when opportunities for employment were available to them: "Only the man who can show proof that he has sought and been refused employment

---

24 See, for example, comments by Pitwood (1940) and Timber (1940).

or engagement is to receive any consideration for aid on departmental account" (Modsell, 1939, p. 4). Notably, at that point in the recruitment process, the NOFU was the only wartime unit in which work was available.

This policy was one of many obstacles to securing relief that residents had encountered during the 1930s (Overton, 2001). There was certainly disapproval for taking the dole. "The country cannot afford shirkers," wrote the editor of *The Evening Telegram*, "those able-bodied men who, if they chose, could be supporting themselves" ("Relieving Officers' Notice," 1939, p. 4). Such comments reflected a belief that people who sought relief were choosing the desperation in which they lived as a viable alternative to work. Such comments also placed pressure on men to enlist or else have their loyalty questioned, as well as their masculinity. What this newspaper editor felt could not be afforded was both economic and socio-political shirking; it concerned both the nation's budget and the national—masculine—image.

The word *shirker*, here, was the civilian equivalent of *slacker*, a dreaded and dreadful word widely used during the First World War to belittle and shame those men who were judged to be avoiding military service.[25] In the comment above, "shirker" was tied to the concept of work. And it is suggestive of how work—as well as the massive unemployment of the 1930s—shaped concepts of masculinity in the interwar period, including during the Great Depression. A man who was without work and failed to provide for his family was often considered an inadequate man (Griffin, 2024). This attitude points to what may have been another factor behind the rush to serve at the onset of war. It was a cruel irony, then, that work in the NOFU became the object of criticism in the UK, and was framed around a notion of masculinity measured by "being visibly a member of the fighting services," else a man could be "given a white feather to mark him as a pacifist and unmanly coward" (Rose, 2003, p. 178).

Three months into the first six-month contract period, a tally was published of the number, by community, of men enlisted in the NOFU and the

25 The word *slacker* accompanied by a white feather—a symbol of cowardice—was used to police gender identity and manhood in the country, and in the British Empire more generally. For a Newfoundland example, see V. S. Walsh (1918). For analyses of the Order of the White Feather campaign, see Gullace (1997) and P. J. Hart (2010).

resulting number of families taken off relief. From the 116 community Relief Offices on the Island that were included, a total of 2,134 men had enlisted in the NOFU, and a total of 392 families had been taken off relief ("Forestry Unit and No. Families," 1940). The threat of being denied access to relief remained a means of encouraging recruitment and retention as the war went on. One year later, on a visit to NOFU camps, Commissioner of Finance John H. Penson warned men that, should they choose to return home at the end of their contract, relief would not be available to them (Penson, 1941b).

Early in the war, before the wartime resurgence in the economy in Newfoundland, the NCG relief program was still supporting tens of thousands of Newfoundland families. The impact of employment in the NOFU made a significant impact on these numbers. In August 1940, Secretary of Natural Resources Claude Fraser presented Commissioner Gorvin with a list of NOFU men and the communities from which they came. He indicated that the distribution of NOFU allotments—the 50 percent of men's wages paid to their families in Newfoundland—was geographically widespread and totalled approximately $100,000 per month (C. Fraser, 1940), an income otherwise not available to families at the time. By the end of the NOFU's time in the UK, the total amount of allotments sent home by members of the unit was $3,636,565.71 (*Annual report*, 1946, p. 41).

Economic relief brought about through the intersection of employment and war service was used as a line of defence for the NOFU in the spring of 1941, when rumours were swirling that another large number of men whose contracts had expired would be returning to Newfoundland. Replying to criticisms of their decision, a columnist from St. Mary's Bay[26] and Salmonier—an area to which many of the men would return—argued they should be praised, not vilified as unpatriotic "money-grabbers." The columnist further challenged

26 Throughout the history of Newfoundland and Labrador, bays were—and to a lesser extent still are—an important geographical and cultural identifier. Throughout this book, the bay in which a community is located is usually included when it identifies a particular collective (for example, men from Trinity Bay), when it clarifies which specific community is referenced when there existed (or exist) more than one of the same name, when it prevents confusion among places with only a minor difference in naming (such as a singular or plural form of the same name), and when it provides general locational information for a community that no longer exists due to resettlement.

the characteristic of dependence cast on the men, which the *Amulree Report* had so vividly portrayed (NRC *Report*, 1933):

> The willingness with which these men came forward in response to the call for volunteers should be applauded. True, most of those who went overseas were unemployed, but a large percentage were not. No pressure was brought to bear on any of them, their service was purely voluntary. Obviously their spirit of independence is above par, since they preferred to risk the danger of enemy action, and deprive themselves of their families, in order to procure for them an independent living in preference to remaining at home, and being among the unemployed.... The Foresters are doing a twofold work—relieving their country of dole recipients and giving their services to the Empire. ("Foresters Not Returning," 1941, p. 6)

Close Calls on the High Seas: Transporting Wartime Workers

The recruitment of workers for the NOFU proceeded hastily but with apparent ease; the transportation of workers overseas, however, was more complicated. A Shipping Advisory Committee, established by the NCG on November 17, 1939, and chaired by Commissioner Gorvin, oversaw the transportation of NOFU workers to the UK and provided advice to the NCG about safely meeting the demands of shipping in wartime. Committee members included representatives from key shipping companies in the country—Furness Withy, Marine Agencies, Harvey and Company, A. H. Murray—as well as the Newfoundland Railway and the Customs Board. Claude Fraser, Secretary of Natural Resources, served as the Secretary for this Committee.[27]

The overseas transport of NOFU workers was the first item of business at the Committee's inaugural meeting. Officer-in-Charge Jack Turner was in

---

27 Claude Fraser was well-known in the woods industry. See "Claude Fraser: Obituary" (1945). A Secretary was the senior public servant in a department of the NCG (Neary, 2011a).

attendance. Due to winter travel conditions and the size of transport ships available, it was initially decided that the most suitable port from which to transport the first contingent was St. John's. To avoid the potential problems related to large numbers of men gathering in the capital city, however, the Committee secured a special train for volunteers; it would collect men at designated points as it ran from west to east across the Island and arrive in the capital just prior to the transport's sailing time (Shipping Advisory Committee, 1939).

The Committee worked expeditiously, making arrangements with the Newfoundland Railway, local shipping agents, and health authorities to ensure the quick and safe preparation of men for travel. The Newfoundland Railway provided both rail and ferry services, including ferry transport between Newfoundland and Nova Scotia, since many of the first-contract workers would travel overseas from Canadian ports—Halifax and Saint John. Locally available ships—Bowring Brothers' SS *Imogene*, Job's SS *Beothic*, and Furness Withy's Red Line ships, the *Fort Townshend* and *Fort Amherst*—were ruled out because they required refit for troop transport and had limited accommodation (Puddester, 1939). Overseas transport was arranged largely through the Cunard-White Star and Canadian Pacific Railway shipping lines. In the early months of war, NOFU workers travelled in small contingents unless they were crossing in a convoy. During the rest of the war, as the Battle of the Atlantic intensified, all were transported in convoy.

Claude Fraser captured the complexity of transporting workers in his August 17, 1940, report to Commissioner Gorvin (C. Fraser, 1940). He outlined the logistics of departure and transport of each contingent, which underscored the level of organization and co-operation involved in mobilizing this workforce so quickly. Details of the departure of the first contingent serve as an example. On December 13, less than a month after the call for workers was issued, the first contingent of three hundred men (led by Officer-in-Charge Turner) departed St. John's aboard the ocean liner RMS *Antonia*. A special Newfoundland Railway train had run from Corner Brook to St. John's on Sunday, December 10, collecting men from across the Island. One hundred men boarded at Corner Brook and Deer Lake, thirty-one at Grand Falls and Windsor, and seventy joined at Lewisporte. Most of this last group had travelled by boat to

Lewisporte from communities around Twillingate Island and area. Thirty-seven men, most from Trinity Bay, boarded at Whitbourne; the remainder arrived in St. John's by truck from Salmonier and St. Mary's Bay. At the Furness Withy Pier in St. John's, they received medical examinations, vaccinations, and tickets for travel. A blanket passport, rather than individual ones, was provided. A tugboat transported men from the pier to the RMS *Antonia*, which was anchored two miles outside The Narrows. The group travelled without convoy and arrived at Liverpool, England, five days later, on December 18 (C. Fraser, 1940).

Other contingents followed; their arrangements were no less complex and also involved travel by train and ferry to reach Canadian departure ports. By mid-winter 1940, just under three months after the initial call for workers, more than two thousand NOFU men were in the UK (C. Fraser, 1940; NOFA, 1945; Turner, 1943a). They had taken with them the tools of their trade: 800 bucksaw frames; 6,132 bucksaw blades; 1,216 axes; 600 axe handles; 2,040 files; 236 peavies[28]; 120 peeling knives; 100 saw sets; 8 jointers; 46 crosscut saws; 84 raker gauges; and 20 clamps (C. Fraser, 1940).

A train carrying NOFU workers departing Botwood for St. John's and overseas transport, 1940. (Courtesy of the Botwood Heritage Society, 9.21.003)

28 A peavey is a long wooden stick with a two-ended hook (one short and pointed, the other long and curved), used to shift logs.

Transatlantic sea travel offered its own dangers in winter, which warfare compounded exponentially. Larry Gladney (NOFU #1901) was one of 963 men who travelled in convoy from Halifax on the Polish liner SS *Chrobry* in late January 1940. In his memoir, he described an incident in which seawater crashed through a broken porthole, and another in which rough seas pushed a convoy ship into the side of the *Chrobry*, shifting its cargo. The men had to rush to rebalance the ship to avoid a dangerous list (Gladney, 1999). Michael Armstrong (NOFU #1382), in the same contingent, recounted that despite these rough seas, very few men got sick. Armstrong also described nightly concerts in the dining room, when men "sang songs, gave recitations and step dances, and played musical instruments" (1940a, p. 1).

Contingents in the later second and third groups faced more perilous conditions when crossing the ocean. After the fall of France in 1940, German U-boats extended their range in the Atlantic, which had a crippling effect on ships delivering essential wartime supplies and personnel to the UK from North America. The last two groups of NOFU workers travelled safely to the UK during this time of heightened maritime action, but not without incident. John Nick Jeddore (NOFU #3361) of Conne River described part of his journey on the SS *Mendoza* as a member of a contingent of 251 NOFU recruits travelling overseas during the summer of 1941. Mid-ocean, the *Mendoza* strayed from its convoy in the fog. When it rejoined it the following day, those aboard saw life-boats and jackets afloat where the convoy had been torpedoed (Jeddore, 2015). Another contingent of 191 NOFU members travelled aboard the Norwegian whaling ship SS *Svend Foyn* the same summer. Christopher Roberts (NOFU #3475) of Sally's Cove, who was 16 at the time, was aboard. He described a collision with an iceberg that left a 10-foot hole amidship—luckily above the water line—as well as seas too fierce to lower lifeboats if it became necessary to abandon ship (C. Roberts, 1998).

Despite the dangers, no details about insurance (should injury or death occur during transport of NOFU workers) had been clarified. This "vexed question" (St. John Chadwick, 1942a, p. 3) was finally resolved in late 1941 through an amendment to the United Kingdom Personal Injuries (Civilians) Act (Dunn, 1941a). However, by then the second and third groups had already travelled to the UK. Despite close calls, the NOFU recorded no conflict-related injuries and deaths during transport between Newfoundland and the UK during the war (T. V. Curran, 1987).

The first NOFU contract provided for free return transportation for workers upon its expiry, as well as for medically discharged members. Once the men were in the UK, however, issues emerged that required further consideration of these stipulations. For example, Captain Turner had flagged the problem of men arriving in the UK who had been sick when they left Newfoundland (Turner, 1940a) and questioned the stringency of the medical examination given to members before their departure. By March 1940, only three months after the men's arrival in the UK, a dozen of them were awaiting repatriation because of pre-existing physical issues. Turner warned that he anticipated even more men would be repatriated on medical grounds over time (Turner, 1940b). Stricter exams before departure from home, he felt, could eliminate some of these cases and decrease unnecessary transportation costs.

Policies would later also be reformed regarding transport of certain groups of men, including workers whose contracts had been completed and those in breach of contract. Timing for the return transportation of time-expired workers was eventually limited to a three-month window after their contract had ended, and men judged to be in breach of contract were required to fund their own return transportation. NOFU members who transferred to the armed forces, whether demobilized or declared unfit, received free transportation to Newfoundland at the cost of the UK government (Dunn, 1942).

The haste with which the NOFU was raised, while an expression of support for the Allied effort, brings into focus other factors that may have motivated the NCG and the men themselves. Expediting the contracts and the transportation of such a large number of mostly rural-based workers took the men and their families off the relief books at home and thereby decreased a burden on the public purse. Such expeditiousness, however, also laid the groundwork for new problems once the unit arrived overseas.

Transportation to UK of NOFU Groups and Contingents[29]

Recruit Group	Departure Date Range	Contingent Size	Contingent Leader	Ship/Line and Vessel Type
One	Late Fall 1939 to Winter 1940	#1–300	NOFU Officer-in-Charge Jack Turner of St. John's	RMS *Antonia*, Cunard–White Star Line (CWS) passenger liner
		#2–304	Bruce Fraser of St. John's	RMS *Antonia*, CWS passenger liner
		#3–409[30]	Thomas Curran of Gambo	SS *Duchess of Richmond*, Canadian Pacific Steamships (CP) passenger liner
		#4–963[31]	Edgar Baird of Gander	MS *Chrobry*, Gdynia Atlantic Line (GAL) passenger liner
		#5–174	Alleyne Hubbard of Grand Falls	SS *Duchess of Richmond*, CP passenger liner
Two	Summer 1940	#1–205	Maurice Job Taylor of St. John's	RMS *Antonia*, CWS passenger liner
		#2–203	Noah Tilley of Robinsons	SS *Duchess of Richmond*, CP passenger liner
		#3–204	Frank Peters of St. John's	RMS *Scythia*, CWS passenger liner
		#4–392	Charles Cahill of St. John's	SS *Ettrick*, Royal Navy troop transport ship
Three	Summer 1941	#1–251[32]	Bela Rendell of Heart's Content	SS *Mendoza*, Ministry of War troop transport ship
		#2–191	Michael Collier of St. Alban's	SS *Svend Foyn*, St. Helier whaling ship

29 This information was collated from C. Fraser (1940) and T. V. Curran (1987), whose contingent numbers do not match. Curran's figures are in the table. He provided no account of contingent leadership for the second and third groups, however; names in the table are from Fraser and other Department of Natural Resources correspondence of the time. There is general discrepancy across sources on the dates of departure and arrival, usually by a day or two.

30 James Carew (NOFU #952) of Cape Broyle and Patrick Carew (NOFU #953) of nearby Admiral's Cove composed a tribute song to this contingent entitled "The Foresters' Song." See Kelly (2014) and Kelly & Forsyth (2018).

Overseas Departure Port and Date	Arrival Port and Date
St. John's, Newfoundland, December 13, 1939	Liverpool, England, December 18, 1939
Halifax, NS, Canada, January 14, 1940	Liverpool, England, January 23, 1940
Bay Bulls, Newfoundland, January 13, 1940	Liverpool, England, January 19, 1940
St. John's, Newfoundland, January 24, 1940	Firth of Clyde, Glasgow, Scotland, February 8, 1940
Saint John, NB, Canada, February 4, 1940	Liverpool, England, February 12, 1940
Botwood, Newfoundland, June 23, 1940	Liverpool, England, July 14, 1940
Montreal, PQ, Canada, July 8, 1940	Liverpool, England, July 19, 1940
Halifax, NS, Canada, July 18, 1940	Liverpool, England, July 24, 1940
St. John's, Newfoundland, July 26, 1940	Firth of Clyde, Glasgow, Scotland, August 17, 1940
Montreal, PQ, Canada, July 7, 1941	Liverpool, England, July 28, 1941
Halifax, NS, Canada, July 19, 1941	Liverpool, England, August 7, 1941

31 Angus Temple (NOFU #1379) wrote a tribute to this contingent, a long narrative poem entitled "Convoy to Scotland." See Temple (2018). See Barrett (1940) for a diary account of this trip.

32 John Nick Jeddore included an account of this voyage in *Moccasin Tracks*. See Jeddore (2015).

SECTION TWO

OVERSEAS EXPERIENCES: WORK, WELFARE, ORGANIZATION, AND EDUCATION

Despite the urgency with which the NOFU was requested, raised, and transported, upon arrival in the UK the men found their hosts unprepared for them. Neither sufficient forest tracts nor wood harvesting and transport equipment had been secured, and an adequate pay system was not in place (Penson, 1941a). Newfoundland's Commissioner of Defence, Lewis E. Emerson, later commented that it was likely that the UK government had not expected the success of the appeal for workers in Newfoundland to have been "so great or so speedy" (Emerson, 1942, p. 4). A huge display of patriotism and less-than-subtle coercion by the NCG combined with pent-up demand for employment and efficient organization by the Department of Natural Resources resulted in men on the ground long before the UK government generally and the Forestry Commission specifically were prepared to receive and assign them any meaningful work.

At the onset of war, the UK Forestry Commission was divided into two branches: Forest Management and Timber Supply. The Timber Supply branch oversaw the assignment of forests tracts for wartime felling. As in the First World War, without a sufficient source of timber from state-owned forests, Timber Supply secured cutting rights for mature forests on private estates, most of them in the Scottish Highlands. But such acquisitions took time. Not many had been identified or cutting rights negotiated when the first NOFU contingent arrived in late 1939 (Duncan, 1942).

These unfortunate delays, along with other aggravating factors (including weather-related delays and unit supervision), made the first six months overseas a challenging time and cast a shadow on the work of the NOFU that became a persistent liability. By the end of 1940, dissatisfaction was so high and relations between the NOFU and the Forestry Commission so poor that consideration was given to dispersing the men or dissolving the unit and repatriating them (Home Grown Timber Production Department, 1941). Commissioner Penson judged these options would be "nothing short of a disaster from the political point of view" (Penson, 1941a, p. 1), as far as relations between the UK and the NCG were concerned, especially around Newfoundland's contribution to the war effort.

These tensions did not abate until the end of 1942. In the meantime, they raised questions about the NCG and UK governments' support for the unit, the attitudes they held about its members, and the degree of credibility and respect they accorded its leaders. They also reflected the extent to which the unit's wartime role was embroiled in wider colonial politics concerning the status of Newfoundland in the prewar period, and wider assumptions about the moral character of its population, such as those that had been conveyed in the *Amulree Report*[1] only a few years earlier.

"Quite a Lot to Learn": The First Six Months

Upon its arrival in the UK in late 1939, the NOFU was attached to Timber Supply, which oversaw production matters related to the unit. The NOFU headquarters was established in Edinburgh with the Scottish Forestry Commission, in anticipation of their work becoming concentrated in Scotland; it included administrative offices for Turner and his second-in-command, General Superintendent Joseph Curran, as well as an accounting branch for the

1 See, for example, James Overton's critique of some of these assumptions in the *Amulree Report* as they relate to poverty in Newfoundland at that time (Overton, 2001).

unit.[2] All matters related to the NOFU passed through the office of Sir Samuel Strang Steel, who directed the Forestry Commission's operations in Scotland, and from there went to the Director of Timber Supply, who was located in Bristol.

Like their Newfoundland Forestry Companies counterparts in the First World War, who had lived under canvas for the first two months of their initial posting (Kelly, 2020), most of the NOFU men also spent their first months in temporary camps, organizing and acclimatizing (Turner, 1943a). Along with the chaotic operational start, the men were also adapting to their new context: a different climate, country, and culture that included changes to their diet; new wartime conditions and restrictions; and new methods and cycles of harvesting and lumbering. They adapted to machinery shortages and different ways of felling, transporting, and milling. It was during this period that relations between the unit and the Forestry Commission were established and first tested.

The NOFU had been initially tasked with cutting round timber for pit props, which were required in varying lengths and sizes. Prior to the war, orders for pit props went to brokers who cut to specification from timber stockpiled in their yards. During the war, orders came from brokers to the forestry units and were filled and shipped directly from the camps. This change added a level of complexity to the work to which the men were unaccustomed. There were approximately 300 different sizes and types of timber to cut, based on a specified relationship of diameter to length (such as 3.5 inches diameter by 4 feet or 3.5 inches diameter by 3.5 feet[3]), each of which had to be shipped to a different firm. It was "complicated business" (J. Curran, 1940, p. 3). The specificity of the work required new skills and patience, as well as qualified oversight.

2 As of February 1940, the operational and administrative offices of the NOFU were at 28 Drumsheugh Gardens, Edinburgh; by May they were at 4 Douglas Crescent, with the Timber Supply Department of the Forestry Commission. In July 1942, several offices—accounts, supply, and administration—were moved again, to 22 Drumsheugh Gardens with the Home Grown Timber Production Department of the Ministry of Supply. By this time, operational headquarters were established at Carrbridge where Officer-in-Charge Turner was stationed.

3 Roughly 8.9 centimetres by 1.2 metres or 8.9 centimetres by 1 metre.

Two NOFU workers preparing pit props, 1941. (Courtesy of The Rooms Provincial Archives Division, A 7–165)

When forest stands were available and approved for harvesting and milling, organizing these tasks under conditions that included adverse weather, a shortage of machinery, and other setbacks, affected the unit's productivity levels. As Turner noted, there was "quite a lot to learn," especially as the work of the NOFU also later expanded to product types beyond pit props. It then also tackled the challenge of producing highly specified lumber while combining "Scots gear and Newfoundland ideas into a mill that would turn out reasonable quantities" (Turner, 1943a, p. 9).

But there was more to the story than what Turner conveyed in his first public report, which would only be published in Newfoundland in early 1943 as a summary of the work of the unit—a full three years after the unit's arrival in the UK and at the end of a serious crisis of well-being and credibility. The tone of that public report—an "all's well with our boys" story—differed markedly from what was written in the monthly confidential reports Turner was required to file for the NCG. Those reports were not public relations material. They were sent first to Timber Supply, where the Director reviewed each one and added a covering letter—which often criticized Turner's account of things. Both missives were then sent to the NCG's Commissioner of Natural Resources, who said little or nothing in response. These reports not only detailed the difficulties the unit was experiencing, they also revealed the differing perspectives and tensions between the Officer-in-Charge and Timber Supply. It was not what the unpopular NCG expected nor wanted to read—that it was overseeing poor treatment of a contentious unit of Newfoundland workers overseas. Fallout was imminent.

As Minister of Supply Sir Andrew Duncan admitted in 1942 in a memo to Secretary of State for Dominion Affairs Clement Attlee, the NOFU had arrived on UK soil much quicker than anticipated, before adequate arrangements had been completed. Extreme winter weather and a lack of designated timber stands delayed cutting operations, leaving men in temporary base camps without work. Duncan felt this situation was a key contributing factor in the decision of many of the first group to leave the unit at the end of their contract. He concluded that "the Unit began under the very worst possible conditions and some of the men in the original draft returned to Newfoundland without having touched a piece of timber" (Duncan, 1942, p. 2). The men had rushed to the UK just before the Christmas season. The delay in beginning work—for some it lasted over three months—undoubtedly affected morale. And the whole situation—the combination of illness, restrictions, criticisms, shortages of raw materials and machinery, and low unit productivity—created a perfect storm of poor first impressions for all concerned.

Settling In: From Base Camps to Log Cabins

The first contingent of three hundred NOFU workers was assigned to small forest tracts on the border between Scotland and England: at Kielder, Northumberland, and Kershopefoot, Cumberland.[4] There they were housed in "base camps," which were Ministry of Labour civilian instructional centres that had been closed before the war. The centres provided training and

4 Cumberland was subsumed into Cumbria in 1974.

housing for chronically unemployed youth during the Great Depression and had operated from 1929 to 1938 (King, 1999).

In addition to the training centres in England, three additional centres in Scotland were also used—the first two at Glenfinart and Lochgilphead (Knapdale), and the third at Glenbranter, all in Argyllshire.[5] These base camps functioned as the initial sites for harvesting operations until new cutting sites could be arranged and "operational camps" built by the workers themselves.

But NOFU contingents arrived more quickly than operational camps could be built, especially given the slow pace at which forest tracts were being secured, creating a lag in mobilization to the work sites. Despite the glowing reviews of the base camps by some members, Joseph Curran commented that, for many reasons including work shortage and its impact on morale, "it was unfortunate that men had to remain for any time in these camps" (J. Curran, 1940, p. 1).

It was of some comfort, however, that the base camps offered greater amenities than those to which most of the NOFU men were accustomed.

The Ministry of Labour Civilian Instructional Centre at Kielder, Northumberland, England, which was used to house the first NOFU contingent, 1940. (Courtesy of the Botwood Heritage Centre, 9.18.071)

5 From 1975, Argyll and Bute. Throughout this text, I use the name of a shire or county that was in use during the historic period under discussion. As here, in such cases, the contemporary name is footnoted.

Larry Gladney, who arrived with the fourth contingent of the first group, was at Glenfinart; based on his experience there, he described the instructional centres as "a wonderful set of camps" (Gladney, 1999, p. 45). Michael Armstrong, with the same contingent, described living conditions as "exceptionally good," highlighting, in particular, a recreation room that had a piano, a billiard room, a gymnasium, and a reading room with radio and newspapers (Armstrong, 1940a, p. 1). But there was one way in which Newfoundland woods camps outranked the UK base camps: warmth. Turner noted that, despite having electricity and modern kitchen and bath facilities, the base camps were felt to be "almost unbearably cold due very largely to the fact that the climate was much damper than our own, and the local idea of a comfortable temperature indoors is about what we would consider rather chilly outdoors" (Turner, 1943a, p. 9). To add to the misery, many NOFU members became sick with respiratory illness.

This early stage of transitioning from base camps to operational camps required the support of locals, who opened their homes to the unit as it set about organizing its work. While operational camps were being prepared,

The Ministry of Labour Civilian Instructional Centre at Glenbranter, Argyllshire, Scotland, featured prefabricated steel Nissen huts with their distinctive curved roofs, 1940. (Courtesy of the Botwood Heritage Centre, 9.18.070)

the men lived in any available accommodation near their cutting site, from family barns to village halls. Larry Gladney described living conditions near Stanley, Perthshire, where he was reassigned from Glenfinart to build operational camps: "There were three small huts for us to stay in. There wasn't a stove, kettle, pan, or light, but plenty of food and blankets and sections of mattresses" (Gladney, 1999, p. 48). The kindness of neighbours and the estate owner transformed the situation, but not before seventeen men came down with diphtheria from a contaminated water source (Gladney, 1999). Elizabeth Ormiston Kelly of Newtonmore, Scotland, described the impact on her life when NOFU workers arrived in her Highland village in February 1940. Her family home, Glenbanchor, became an office for NOFU operations at Laggan, where the unit built six camps and a sawmill; her family relocated to a house in Newtonmore (Ormiston Kelly, 2002).

With no treated lumber available, the NOFU used the raw timber they were cutting for the war effort to construct their own operational camps. The men were used to building their own camps—that was part of their logging work in Newfoundland—and they made swift work of establishing themselves in the Highland forests using a "snowball process" (Turner, 1943a, p. 9). They constructed a twenty-man bunkhouse in two or three days, usually, and then moved in a crew of men. Then the next one was constructed, and more men were added. The process continued until a camp was complete, with bunkhouses, cookhouse, forepeak, etc.[6] Camp size was determined by the number of loggers needed to fell the nearby forest allotment. Turner explained the template for camp construction:

> A special small gang … does all the construction, including plumbing and water supply. As the number of men in our newer camps varies from 80 to 120, there must be some variation in the size of the camps.… [W]e use a standard bunkhouse, holding twelve men, and merely vary the number of bunkhouses. The cookhouse, which consists of dining room, kitchen, pantry, and

6 The forepeak was the camp building that housed the foreman, second hand (or sub-foreman), and any visiting officials.

> cooks' living quarters, is built in the shape of a capital "T," with the kitchen forming the stem. The width of both parts of the building is the same in all cases, so for a bigger camp we only have to lengthen the two parts. We use a standard ablution hut in each camp. As these are all alike, it is not much of a job to move the plumbing from a cut out camp to a new one. . . . Forepeak, stores, etc., are the same in all the newer camps. (Turner, 1943a, p. 9)

In late 1942, recreation huts became a standard feature of the camps; they were built in the same way as the cookhouses, with an annex added for the canteen and the manager's living quarters.

By spring 1940, sufficient numbers of operational camp sites had been established and, when allotted timber sites near the base camps had been fully harvested, these camps were closed (Turner, 1943a). Excluding the five base camps, the NOFU operated a total of 71 camps at various times between 1940 and 1946. With the exception of two in the New Forest of southern England, all were in Scotland (T. V. Curran, 1987).[7]

Kildrummy camp, Aberdeenshire, 1940. The style of the Kildrummy camp was typical of the camps built by the NOFU. (Courtesy of the Ballater Historic Forestry Project Association)

7 For a list of camps and years of operation, see T. V. Curran (1987).

The NOFU camps became part of the landscape of wartime in the Highlands and their novelty was a point of interest for the UK press, who were encouraged by officials to build goodwill for the NOFU through positive press coverage, especially during the difficult first years of their term overseas. Describing the camps for readers, one newspaper correspondent wrote:

> It takes these men only two days to make one big cabin. They cut down a number of trees to make a site for their home and then, cutting off lengths from the trees they build their cabin. To keep out the cold winds they pack moss between the timbers, and hope that when summer comes, this moss will also keep out the midges. ("Lumberjacks in Scotland," 1940, p. 6)

W. J. Passingham visited a NOFU camp near Ballater in the winter of 1941 to report on the NOFU for readers of *Illustrated* magazine. In his cover story, he concurred with Turner's observations about one superior feature of the camps:

> There are log huts built by the Newfoundlanders, set in a forest clearing and exposed to the bitter weather on mountain sides, which for warmth and comfort surpass anything suburban builders have produced.... [S]pending a few minutes inside them from the bitter weather one realizes that English and Scots alike have not yet learned how to keep themselves warm. (Passingham, 1941, p. 3)

In the second year of operations, prefabricated structures became available, which increased the ease of camp construction and allowed for hut sections to be moved to other sites when a camp closed. These structures were also used to build the NOFU operational headquarters at Carrbridge, Inverness-shire, which opened in late 1941 (T. V. Curran, 1987).

First Impressions: The UK Press and the NOFU

The initial press coverage introduced the NOFU to the British public through feature articles laden with caricatured impressions of the men. Later, news articles reported on some men's behaviour outside the camps and in nearby communities, which fuelled largely negative stereotypes. The publicity conveyed some of the meaning of their work and their identities, but it also fed public and political debate about the state of Newfoundland and its status, which was likely never far from the minds of UK politicians and the members of the public who followed such issues.

With the exception of only a very few members, some of whom had served in the Newfoundland Forestry Companies in the First World War, most NOFU men were crossing the Atlantic for the first time when they went to work in the UK. The land and culture they found there were, in many ways, new to them—and the civilian dress of the men was a novelty to the people of the "mother country." W. Rawdon Smith of the Forestry Commission met the first contingent of loggers when they disembarked in England on December 18, 1939. He described them as "a fine body of men—not the giant lumbermen of fiction,

A member of the NOFU sharpening a bucksaw blade, 1940. (Courtesy of The Rooms Provincial Archives Division, A 8–5)

but young, keen and wiry in build" ("Newfoundland Lumberjacks," 1939, p. 5). A reporter commented on the "brawny timbermen . . . wearing their familiar jerkins, laced high boots and caps" ("Loggers To Work Here," 1939, p. 6).

Nevertheless, the UK press was more inclined to describe the men in less realistic, almost mythic terms. Some writers drew directly on fiction for their accounts, conjuring images of Newfoundland "hillbillies" living in the Scottish backwoods; after a hard day's work, they would temper their voracious appetites with a huge meal and then gather around the bunkhouse stove where "pleasant drawling voices sing backwoods ballads that echo way up amongst the pines—a scene from a Gene Autry film" ("Hill-billies," 1940, p. 19). The theme was echoed by other writers, as in this account of the men of Kielder camp on a day off:

> Newfoundland lumberjacks who came to Hawick by special train from Kielder on Saturday lived up to the pictures of them conjured up for us in the "Wild West" novels—lithe, lanky, powerful men in picturesque garb, and seemingly capable, if the spirit led them, of shooting up the old burg and wrecking the saloon with six-shooters. But, of course, they did no such thing, being for the most part douce and modest fellows, very much intrigued with the Hawick dialect, and quite unconscious of their own strange English, flavoured with French intonations. ("Lumberjacks," 1940, p. 4)

Another journalist commented on the "intriguing comedies" that arose when NOFU men and Border locals attempted to communicate, each with their own richly accentuated English—"as near an approach to Babel as you could wish to hear" ("Hill-billies," 1940, p. 19).

Angus Temple of Sunnyside, who travelled on the SS *Chobry* with the fourth contingent of the first group, noted the media's penchant for a Newfoundlander who fitted a preconceived image. In his poem "Convoy to Scotland," Temple humorously described arriving at the Gourock docks on February 8, 1940, where the press awaited them and the men scurried for attention:

Before we left the dockside
Some newsmen from the press
Came down to snap the lumberjacks
The biggest and the best.
We had some hefty fellows
Among our gallant crew
We all tried hard to get a place
But they said ten will do.
(Temple, 2018, p. 348)

A correspondent for the *News Chronicle* provided a more in-depth picture of the men of the first base camp at Kielder, just south of the Scottish border, but it, too, was in no less mythic terms. Pollard described "young giants" who "sing and dance in Britain's cold backwoods," one "like a great boy who had never grown up" (Pollard, 1940, pp. 1, 6)—perhaps an unintended imperial metaphor for colonial Newfoundland. The image of the "hardy and loyal" (Yarham, 1940) men of the "oldest colony"—or the "Ancient colony" of the "Mother Country" or "Old Country"—conjured a paternalistic relationship. As Peter Neary points out, along with this mythology emerged one in which Newfoundlanders and Labradorians were "children of a harsh but compelling land and the possessors of a rough-hewn but unique cultural heritage which it was their duty to preserve and transmit" (Neary, 1969, p. 43).

The NOFU members were seen as men of "fog and fish," "tough, hard-bitten, and liberty-loving," for whom "the perils of war are hardly so great as those which [they] daily face as part of their ordinary routine" (Yarham, 1940, p. 1). While acknowledging the hardships of work and life at home, Yarham failed to mention that the liberty-loving men of whom he wrote, under rule by the NCG, had no real political liberty. Such ubiquitous, selective, and hyperbolic images announced a colonial identity in which being a loyal subject trumped being a critical citizen. Embedded in their prose and positioning was a specific notion of the Newfoundland colonial man and colonial masculinity: a caricature of physical strength and bravery, largely devoid of intellect or reason, and ever loyal to empire. These images appeared not only in the press but in songs, stories, and recruitment campaigns throughout both wars.[8]

For many Newfoundlanders and Labradorians, the meaning of images of colonial bravery and strength was likely shaped by how the most tragic events of the First World War had been publicly mediated, particularly the Battle of Beaumont-Hamel, at which the First Battalion of the Newfoundland Regiment had been almost annihilated (Harding, 2001). Lord Northcliffe, founder of the Anglo-Newfoundland Development Company, had recalled this imagery and meaning in an address during the First World War to the Newfoundland Forestry Companies at Kenmore, Scotland, in 1918: "[T]he claim of Newfoundland to be the most purely British overseas community in the world is probably accurate. None of the King's far-flung island peoples rushed to his standard with greater zeal and rapidity. None have exceeded the valour of the Newfoundland troops who fought at Beaumont-Hamel" ("Address," 1918, p. 3). His words recalled those of Newfoundland's Prime Minister Morris at the first Memorial Day in 1917: "Those who died that day [at Beaumont-Hamel, France] were heroes. They represented the best of Colonial life . . . Nothing in the country today is as valuable as their memory" (Morris, 1917b, p. 6).

These troubling messages took hold and continued to circulate in the decades following the First World War.[9] As their songs and stories suggest, those lost in First World War conflicts were never far from the minds of the men who went overseas in the Second World War.[10] Their memory was part of what had inspired many of them to volunteer in whatever capacity they

8 One such image—that of a plain-clothed woods worker supposedly "somewhere in the UK"—was used to pitch a UK tobacco advertisement; the copy suggested that if, like the "hefty lumberjack, you've a real man's job to do . . . you'll want a man's tobacco"—in this case, Ogden's St. Bruno ("The Man in a Man's Job," 1940, p. 7). In Newfoundland, meanwhile, front-page newspaper advertisements reminded the public that "The boys overseas prefer Wings. Send a carton today" ("The Boys Overseas," 1940, p. 1).

9 For a discussion of the politics of this messaging and "the myth of Beaumont-Hamel" that emerged from it, see Cadigan (2009). See also Cadigan (2016) for a discussion of "the cult of the fallen" and its historical impact.

10 For a summary of Newfoundland's participation in the First World War, see M. Baker (2017).

could. One song that bridged the efforts of those who fought in the First and Second World Wars and underscored this daring masculinity was oft sung in the NOFU bunkhouses and mentioned in articles about their work. Sung to the tune of "The Wearing of the Green," only a few verses remain. In them, Newfoundlanders and Labradorians who fought in the Great War and the sealers of the SS *Ranger*[11] are offered as a means to counter Hitler's advance through Europe:

> Over there in '14
> The officers in command
> Could not help but praise them up
> The boys from Newfoundland.
>
> Hitler needs a hundred-thousand men
> And Chamberlain takes none.[12]
> I'd take the crew of the *Ranger*
> And the battle would be done.
>
> First he marched on Austry[13]
> Led by a German band
> You'd bet he'd have a harder time
> If he came to Newfoundland.
> ("Lumberjacks—The Newfoundlanders," 1941, p. 7)

---

11 The men of the SS *Ranger*, like all sealers, were often saluted for their courage at work in perilous conditions on the ice floes off the northeastern coast of Newfoundland, which they called "the Front." The 211 sealers who signed on to this ship for the 1902 spring seal hunt were known for another kind of courage—leading a protest march of sealers from St. John's harbour to Government House on March 8 that year, to challenge the rumoured low price that had been set for seal fat (at the time the most valuable part of the animal). It was the first organized job action by Newfoundland sealers (Busch, 1984).

12 Neville Chamberlain was Prime Minister of the UK at the outbreak of war in 1939 and was widely criticized for his policy of appeasement toward Hitler and Germany leading up to it.

13 Austria.

Much of the interest the press showed in the NOFU men was not a response to the men's altruism nor attributable to the novelty of their presence, position, or appearance. Media reporting was at the beck and call of government; it was summoned by officials when needed to "press" a message on its behalf. When a mass exodus of NOFU workers was anticipated in summer of 1940, for example, both Turner and the Forestry Commission agreed that "some appreciative publicity" would be useful to encourage NOFU members to remain in the unit (Ewbank, 1940b, p. 2). As a welfare crisis threatened an additional loss of NOFU workers in 1942, "more propaganda and publicity" in both Newfoundland and the UK was suggested to help remedy the situation (St. John Chadwick, 1942c). A major article on the unit by A. E. Turran published in *The Geographical Magazine* was written under that pseudonym by St. John Chadwick as a fundraiser for welfare purchases (St. John Chadwick, 1942f); a secondary purpose, no doubt, was to enhance the profile of the unit at a critical time in its work term.[14]

By 1941, there were stories in the press that also revealed some of the fracture lines in these tropes of identity. They suggested another side of colonial masculinity, an emerging reputation of the "wild colonial boy." It, too, threatened to become a liability for the unit, as bailies issued fines, warnings, and advice to NOFU workers on trial before them for unbecoming behaviour.[15] For example, a fracas at a hotel in Dun, in which four NOFU members assaulted a bartender whom they claimed refused to serve them (apparently proclaiming, "You are lumberjacks. Get out!") resulted in several charges and this admonishment from the bailie: "I am sorry to see British subjects from Newfoundland in the dock of this court. . . . You have been quite well-treated in this part of the country. . . . Keep clear of trouble in future" ("Lumberjacks Fined," 1940, p. 3).[16]

14 The pseudonym may have been a combination of the first initials of the unit's administrative assistant, A. E. Hart, and a surname that is a portmanteau of Thomas Curran, NOFU Superintendent.

15 In this context, a bailie is an officer of the court; a magistrate.

16 The construction of this particular image of colonial masculinity—the Newfoundland civilian working-class male—during the Second World War is not yet adequately explored as a contribution to the literature of gender and war.

At court in Kirriemuir, Angus, in the fall of 1941, two NOFU members faced breach-of-peace charges and a third faced the same charge plus resisting arrest and interference with police. The unit itself received a warning from the bench in the proceedings. With Officer-in-Charge Turner present, the bailie commented on the increase in such occurrences which, if continued, would result in jail sentences in lieu of fines ("Disorderly Lumbermen," 1941). Earlier that fall, also at Kirriemuir, six NOFU workers had been arraigned on the same charges for creating what the bailie called "disgraceful scenes" and which the press reported as a "mêlée" ("Lumbermen in Court," 1941, p. 1).[17]

The increasingly negative public impression of the NOFU, then, intersected with the unit's political challenges. In the court of public opinion, the impact of what officials called "undesirables" (Viscount Cranborne, 1941a, p. 3) or "troublemakers" (Ewbank, 1940c, p. 1) was felt throughout the unit's time in the UK. One war bride both confirmed the problem and challenged the stereotype: "The general opinion of [the members of the NOFU] wasn't very good because quite a lot of them used to spend quite a bit of their time in the pubs. But I met my husband then, and he was one of them, but he wasn't one of them that went to the pubs" (Stapleton, 1981, p. 23).

Identity Markers: Symbols of Allegiance

Against the stories about them emerging in the UK, the men of the NOFU took pains to style and express what they themselves considered key aspects of their identity. One war correspondent commented that the men of the NOFU "brought with them their own tools, language, songs and, as far as possible, their customs and habits. What they haven't been able to bring are their own climate, food, and tobacco" ("Lumbermen Like," 1941, p. 6). The men

17 See also "Too Many Drinks" (1940) and "Lumberman Fined £10" (1940). Local newspapers in the UK reported instances of charges against NOFU men that included breach of peace, drink driving, drunk and disorderly conduct, bigamy, assault, and obstruction of police, among other misdemeanours. In one instance a superintendent was fined for motor vehicle abandonment that resulted in an accident ("Left Car in Dark," 1941).

good-naturedly complained about the weather and did eventually get supplies of codfish and tobacco. Their tools, language, songs, customs, etc. were, of course, distinguishing aspects of their work and identity. But they also displayed additional cultural symbols.

During its time in the UK, the NOFU was the only civilian unit whose members were not granted a uniform. The Women's Timber Corps and the British Honduras Forestry Unit, also civilian forestry units, were issued such clothing. Significant cultural symbols—the caribou head, the national anthem, and the Newfoundland flag—provided meaningful identity markers for the unit, especially when it came to distinguishing between the NOFU men and those of the Canadian Forestry Corps, who worked nearby: "A lot of the people here seem to think we are Canadians, and although they know we are from Newfoundland, they seem to think we are a part of Canada. Even some of the papers here when publishing pictures of us sometimes call us Canadians" (Armstrong, 1940b, p. 3).

There were the differences casually noted by Ian Cameron: "In 1940s Ballater, we talked about the 'Canucks' and 'Newfies.' To us, they were two very different races. The Canucks had jeeps, big trucks, and crawler tractors and many were French speaking. The Newfies had bicycles, horses, and bushman saws and sounded Irish" (Cameron, 2018, p. 404).[18] Cameron's observations could have easily been a description of the haves and the have-nots of the North American dominions.

The Official Badge of the NOFU

Early in the summer of 1940, as the first six-month contracts were expiring and unit morale was waning, UK newspapers widely reported that a lapel badge was to be issued to all NOFU members. The badge was one of many suggestions to alleviate the discontentment among the NOFU men that they had expressed in discussions with the Duke of Devonshire, Parliamentary Under-Secretary for the Dominions, who had visited NOFU camps in April 1940. The lack of

18 The term "Newfie" is widely attributed to American and Canadian soldiers of the Second World War, who coined it while serving in Britain (Story et al, 2006b). Scots may have learned it from the soldiers of the Canadian Forestry Corps. Many consider the term derogatory. See, for example, Walsh & Russell (1992).

recognition of their work within the UK and disparagement of their status—civilians from Newfoundland—had eroded the men's commitment. A badge that proclaimed their identity was seen as a way to educate the public about who they were and the purpose of their work (Duke of Devonshire, 1940).

Given the pending expiration of contracts, the Dominions Office pursued the matter with urgency. Anthony Eden, Secretary of State for Dominion Affairs, proposed a unit badge to Governor Walwyn in spring of 1940. He emphasized the "National Service aspect" of the NOFU's work in the context of feared large-scale resignation of workers in the coming summer, a matter he regarded "as one of great urgency from [the] point of view of retaining services of men after expiration of original term of contract" (Eden, 1940). There was some back-and-forth between the Dominions Office and the NCG regarding the wording on the proposed badge. Governor Walwyn suggested it bear the inscription "NFLD Loggers Unit," as the men referred to themselves as loggers (Walwyn, 1940). But this phrasing did not resonate in the UK, nor was it the official name of the unit.

The agreed upon final design was a round, silver-plated, buttonhole badge styled with a caribou head at the centre and the unit's official name—"Newfoundland Forestry Unit"—embossed on the border surrounding it (Gaunt, 1940; "Newfoundland Loggers' Badge," 1940). Three thousand were supplied by J. R. Gaunt and Son of Birmingham at 4 pence each and paid for by the Forestry Commission. The woodland caribou, a symbol of the Dominion, tied the unit to the Newfoundland Regiment and its Newfoundland Forestry Companies, which had also used it as their symbol. The badge was an emblem of belonging for the NOFU. It also had practical

The official badge of the Newfoundland Overseas Forestry Unit; this one belonged to Valentine O'Keefe (NOFU #0972), Ferryland. (Photo by author)

uses—in the absence of unit uniforms, members presented it for admission to events specifically organized for members of the Allied services.

The Newfoundland Red Ensign

The Newfoundland Red Ensign became the official flag flown by civilian ships of Newfoundland in 1904. It has a scarlet red background and the Union Jack in the upper left quadrant. The Great Seal of Newfoundland is centred in the right half of the flag and bears the motto *Haec Tibi Dona Fero,* Latin for "These gifts I bring to thee."[19] The Red Ensign was flown not only on ship masts but on community buildings and homes and at public events. At many NOFU camps in England and Scotland, the men flew the Red Ensign on poles that were sometimes topped with a woodland caribou head ("Nfld. Forestry Unit at Work," 1940). It was also used during funeral processions, draping the coffins of members who died and were buried overseas. The Great Seal motto was inscribed on the official NOFU stationery and cards.

This flag, however, was not readily available in the UK. During the Duke of Devonshire's spring 1940 visit, the men raised the issue of flags for the camps—requesting either the Red Ensign or the Union Jack, or both. The men informed the Duke that "in Newfoundland the flying of flags is almost universal, and that no house is without its flagstaff" (Duke of Devonshire, 1940, p. 2). Two years later, in the summer of 1942, the Dominions Office and the Ministry of Works purchased twenty-four Newfoundland flags to loan to the NOFU for the duration of the war. The Dominions Office added the caveat that the unit "take steps to see that they are kept in good condition" (St. John Chadwick, 1942d, p. 2).

The Newfoundland National Anthem

The "Ode to Newfoundland," a poem written in 1902 by Governor Sir Cavendish Boyle and set to music by Sir Hubert Parry, was adopted as the national

---

19 The Great Seal adorned the casket of Newfoundland's Unknown Soldier, who died in battle during the First World War and whose remains were repatriated in May 2024, during the period of his lying in state. At the entombment ceremony at the National War Memorial in St. John's, held on July 1, 2024, a folded Red Ensign sat atop the framed seal. The casket was draped in the flag of Canada.

An illustration of a NOFU camp flagstaff ca. 1940, topped by
a caribou head and flying the Newfoundland Red Ensign.
(Illustration by Jennifer Morgan, 2024)

anthem of Newfoundland in 1904.[20] The popular anthem was sung at official and unofficial functions. During the First World War, recruits from Newfoundland sang the anthem in a variety of contexts, including as a way to distinguish themselves from Canadians (E. Roberts, 2014).

In the UK, much was made of NOFU members' penchant for singing. In the many accounts of camp life, few failed to mention the evening bunkhouse singing and storytelling in which the men engaged to entertain themselves and maintain morale. The "Ode to Newfoundland" often ended such events, as in this instance described by news correspondent George Pollard about the closeout to an evening of bunkhouse entertainment—songs, tunes, recitals, and dances—in the Cheviot Hills of Northumberland:

> Suddenly a word from Captain Turner, and silence. "Lads, lights out soon. Are you ready?" The crowd surged to the piano, and they sang their own anthem, which Cavendish Boyle wrote: "When sun rays crown thy pine clad hills, / And summer spreads her hand / When silvern voices tune thy rills / We love thee smiling land," after which they sang another anthem with equal fervour—"God Save the King." (Pollard, 1940, p. 6)

As the two songs in this finale suggest, their identity was distinctive but also part of the whole—even as the details of exactly how this was so were to be challenged by war and its aftermath, as conditions in Newfoundland changed and the question of its political future and place in the empire intensified.

These cultural symbols—the badge, the flag, and the anthem—were not only important as distinguishing markers of a particular identity. Without uniforms—the standard and idealized masculine expression of wartime identity—the men were challenged to represent the cohesiveness that was an identifiable part of their contribution to the war effort, or a way of expressing the "togetherness and a shared culture" (Robb & Pattinson, 2018) that created

20 The province officially adopted "The Ode to Newfoundland" as its anthem in 1980.

a sense not only of belonging to nation and empire, but of shared purpose and allegiance, too.

Restrictions, Rations, Raids: Wartime in the UK

Shortly after the unit arrived overseas, the UK government introduced large-scale wartime rations at home and in its colonies as a means to manage distribution of scarce resources; the materials rationed ranged from food stuffs to energy resources and clothing. Scarcity was not new to most of the men in the NOFU. They had lived in or through two decades of widespread deprivation, poverty, and disease in their country, which, while unevenly distributed among the members, nonetheless had impressed upon all the hardships of food insecurity, supply limits, and restrictions. Many NOFU members and their families had experienced first-hand the meagre rations of "dole"—the government relief program that had prescribed amounts of certain foods to individuals and families during the Great Depression. Newspapers regularly carried letters of distress from families and community organizations protesting its limits and describing the impact of poverty on the people of the country (Hickey & Puddester, 1938; McGrath et al., 1932).

As well as facing adjustments in terms of the availability of certain foods and clothing while in the UK—including those with cultural meaning—the NOFU workers had to live within the strict imposition of rations and a coupons and points system for purchases. Although some adjustments were made for the men because of the demands of their work, the unit was affected by all rationing, including petrol, which was rationed from the beginning of the war. Clothing rations came into effect in 1941 and were factored into the discussion of unit uniforms. Like the general population, the unit was "on the rations"—and the conflict that made them necessary was never far away.

Food

Food items—including sugars, fats, meats, cheeses, eggs, tea, and bread—were rationed. Although conditions at woods camps in Newfoundland were difficult, a mainstay of comfort was food, which was prepared by talented cooks who kept men well fed and satisfied—and on the job. In his report on conditions in woods camps in the early 1930s, Gordon Bradley had written of many difficulties, but, despite these, he conceded that "the men are substantially fed with food of good quality" (Bradley, 1934, as cited in Neary, 1985, p. 201), and the men could eat as much as they wanted. Despite some allowance for the extra needs of the NOFU men because of the demands of their work, wartime rationing meant there was a limited supply of some foods or key ingredients for cooking and baking camp food, or none at all. John Nick Jeddore did a stint as camp cook and commented: "Rations were so thin it was a job to feed hungry men. You only had one shot at preparing a meal, no do-overs" (Jeddore, 2015, p. 121).

Officer-in-Charge Turner explained the central role of cooks in camp structure, noting that "a forestry unit marches on its stomach" (Turner, 1943a, p. 9) and that in a wartime context, having satisfied men depended on having creative cooks:

> Our cooks had to get down to using many things they had never used before and getting along without many things they'd been using all their lives . . . they had to feed gangs of hungry men without salt beef or pork, beans, molasses, salt fish or dried fruit. . . . In spite of rationing and difficulties of distribution, there is no shortage of good, well-cooked food in our camps. Many of our cooks have performed miracles in making Newfoundland dishes from British materials. When a cook can produce first-class baked beans without either molasses or fat pork, and turn ships biscuit and Iceland fish into really good fish and brewis, he is GOOD. (Turner, 1943a, p. 9)

NOFU members supplemented their diet with wild game, which was not rationed (T. V. Curran, 1987), just as they did at home; so had the Newfoundland Forestry Companies during the First World War. As Turner's comment indicated, they were also able to access dried cod. One newspaper correspondent writing on the food preferences of the various foreign units present in the UK noted that the NOFU, who "are doing a really tough job over here, have a passion for salted cod" ("Army Makes," 1941, p. 7). Icelandic cod, dried and salted at sea and landed at northern UK ports, was rationed to the NOFU at an average of two ounces per day. The correspondent added: "Last week I saw some of it in the shops of London—the habit has spread to civilians" ("Army Makes," 1941, p. 7).

Cooks were in high demand for the unit and the appliances for cooking meals were as important as the tools for cutting timber. In the summer of 1940, the Department of Natural Resources issued a call for cooks at a monthly rate of $75 (C. Fraser, 1940). Earlier that year, when the NOFU was busy building its operational camps, the United Nail and Foundry Company of St. John's completed an order for twenty wood-burning cooking stoves for the unit. The stoves were of the sort used in the camps of the ANDCO and Bowater in Newfoundland. They were made of cast iron and sheet medal, measured forty-four inches long by twenty-four inches wide (roughly 112 by 61 centimetres) with an oven and firebox on the bottom ("Local Made Stoves," 1940).

Clothing

Clothing allowances also affected the unit. In the UK, the men could not avail of many of their preferred clothing items for the civilian work outfit or unofficial "uniform" of the logger. Gerald W. St. John Chadwick of the Dominions Office, writing under the pseudonym A. E. Turran, mentioned some of these items in his description of the NOFU men he observed, mid-war, at work in a camp in the Highlands:

> He wears a pair of logans, stout boots with thick rubber soles and soft leather uppers—just about the most warm and comfortable form of boots one could ever wish to walk in. Above this he will wear his oldest toughest breeches or trousers and an old shirt or sweater: it is not a paying game to waste new clothes which may tear at any moment on protruding twigs and branches. In cold weather he puts on a light wind-breaker, characteristic Newfoundland garment in wool of drab blue-and-brown checks. (Turran, 1943, p. 533)

There were other common elements of the attire—woollen caps and ear muffs, for example—although it was reported that, in response to warmer conditions than they were used to at home, many men had joined "the No Hat Brigade" since arriving in the UK ("Britain's Need," 1940, p. 6). A steady supply of such clothing items was sent on consignment to the NOFU through the Department of Natural Resources. At the end of 1941, for example—a time at which there were an estimated 1,876 men on the ground in the UK—the following order was approved: one thousand windbreakers at $3.30 each, one thousand pairs of breeches at $3.80 per pair, both shipped from Québec City, and one thousand pairs of logans at $4.25 per pair, shipped from St. John's (Walwyn, 1941c).

But there seemed to be no end to the drama of obtaining appropriate wear for the unit. For example, said logans became an item of contention when the Canadian supplier shipped larrigans, the Canadian Forestry Corps equivalent but *not* the boot the NCG ordered or that the men wanted. When settling accounts after the war, the question was raised about whose responsibility it was to cover the misspent $4,000—the UK Ministry of Supply or the NCG Department of Supply (the supplier was absolved since they had simply filled an order). The Department of Justice of the NCG determined that the error had been a "breach of duty on the part of the agent [the NCG Department of Supply] to his principal [the Ministry of Supply]" and recommended accepting the Ministry's offer to bear half the cost (Powell, 1947).

Obtaining off-duty wear also presented challenges. The men preferred wide-legged trousers to the narrow-legged style that was standard in the UK. In 1943, not long after the unit's request for uniforms had been denied, Turner sent an order for such trousers to a Glasgow tailor. That business, in turn, rightly sought an exemption to produce them from the Board of Trade, under the Restriction of Style Orders. Wide-legged trousers exceeded the maximum width allowed and required more fabric. Along with turned-up cuffs, fasteners, elastic, and double-breasted suits, they had been disallowed under the Civilian Clothing Order of 1942 (McNeil, 1993). When the exemption request reached the Board, officials there misunderstood Turner's request, thinking the NOFU Officer-in-Charge was trying to buy unit uniforms without permission. A chain of correspondence followed before the matter was clarified, which ended with one clearly frazzled clerk of the Board of Trade concluding: "I take it that the Newfoundlanders are reconciled to having no trousers" (Jenkinson, 1943, p. 1).

As in any rationing situation, an underground market emerged that targeted the NOFU. Ian Cameron recalled a "new breed of entrepreneur" who arose to meet certain needs of the men from Glenmuick camp near his home of Ballater, Scotland: "The first thing the young loggers wanted was to be decently dressed for Saturday nights out. Despite the wartime clothes rationing, a small car laden with suits made the rounds of the logging camps. Trade was brisk" (Cameron, 2008, p. 24). The men were clearly invested in their style and appearance—a mode of masculine expression suitable to the times and that differentiated work and play—as their response to clothes rationing in the context of new courtship opportunities demonstrates.

Petrol

The strict rationing of petrol combined with the isolation of many of the NOFU camps created hardship for workers. Recreation huts and canteens did not appear in most camps until late 1942, so the only reprieve from work and camp life in the early war years was in nearby villages and towns. The men also wanted access to churches for weekly worship. A supply of petrol for weekly recreational trips to nearby places that offered amenities was available to some forestry units, including the Canadian Forestry Corps. This petrol allowance did not become available to the NOFU until midway through the war, and then only after a major crisis in the unit's status.

The shortage of petrol accounted in part for the popularity of bicycles in the camps. As Ian Cameron pointed out, along with a suit, a bike became an after-work necessity to enable men to get to town:

> The next big purchase on our colonial's mind was a bicycle to get to the dance wearing his new suit. . . . [B]icycles were rationed at two per shop per month. There were no restrictions, however, on bicycle parts and our happy bicycle agent was soon mass-producing cycles in his back shop to try to keep up with the demand. (Cameron, 2008, p. 24)

Two NOFU members ready to cycle to town, 1940, Scotland. (Courtesy of Rhonda O'Keefe Arsenault)

Despite the NOFU men having little familiarity with bicycles—they were not in common use at home, since Newfoundland had no established road network and travel was by sea or rail—the distances to towns and villages, combined with rations on petrol, inspired wide adoption of this mode of travel.

All these war restrictions and rations were woven into every aspect of everyday life. And although the NOFU men were not in the throes of combat, neither was conflict far from them. Blackouts, bombings, sentries, and shootings were also part of wartime work and life, even in the Highlands of Scotland. The few existing accounts of the NOFU's work in Scotland also mention this backdrop of war. The downing of a plane near Camp 6 at Laggan killed several Canadian soldiers (Jeddore, 2015). Axis bombers targeted a textile mill near the Taymount camp (B. Taylor, n.d.). NOFU men extinguished fires caused by German bombings at the New Forest, near the key port of Southampton and the English Channel (Stringer, 2011). The Canadian Forestry Corps also had close calls with enemy fire while posted in Scotland, including during the bombing of Aberdeen in April 1943 (Wonders, 1987, 1991).

The death of NOFU camp foreman Maxwell Hawkins (NOFU #1480) of Glovertown early in the unit's stay in the UK underscored wartime dangers. On Sunday, July 14, 1940, a group of seven NOFU men, Hawkins among them, travelled by private taxi from Carrbridge to Grantown on a weekend outing. It was raining on their late-night return trip and visibility was poor. Approaching a checkpoint, the taxi driver neither saw the sentry nor heard his shouts to halt until he had driven by. As he pulled to a stop, a gunshot pierced the rear car windshield, fatally injuring Hawkins in the head (Hubbard, 1940;

"Lumberjack Shot," 1940).[21] A charge of culpable homicide against the sentry, Francis Lionel Harrington, was withdrawn when the evidence was judged insufficient to prove either that, as rumoured, he was "under the influence of drink" or that, given his duties, his actions were "culpable and reckless and without reasonable excuse" (James, 1940, p. 1).[22]

Contentious Presence: More Than Just a Warm Welcome

As has already been demonstrated, the presence of the unit in the UK became more contentious over the term of the men's first contract and during their initial introduction to the people of England and Scotland. While there was often a positive picture of the NOFU presented in most UK press accounts, the unit also had many detractors. Along with other countries worldwide, the UK had experienced massive unemployment in the decades between the two world wars, and those numbers remained high at the onset of the Second World War. Even with eligible men enrolled in the armed services or other wartime work, by 1940 estimates suggested around one million people were unemployed (Reubens, 1945). Against this early war backdrop, many UK residents were not happy about the arrival of a large foreign workforce. The Ministry of Labour was criticized for its failure to anticipate the need for forestry workers and train them, which would have forestalled the use of migrant workers ("More Lumberjacks," 1940).

Announcing the arrival of the first contingent of NOFU members, *The Northern Whig* assured its readers that these Newfoundland workers "will not be detrimental to British labour, in fact, their employment will mean more work for unskilled workers here—for example in the loading and transport of timber and other work in the saw pits" ("Loggers To Work Here," 1939, p. 6). The Ministry of Labour and the Forestry Commission hoped to integrate unskilled and skilled workers for useful employment and experiential

21 For descriptions of the event and a list of the men who were with Hawkins, see Hubbard (1940).

22 See also "Homicide Charge Deserted" (1940).

training ("2000 Lumberjacks," 1940). Not everyone was convinced, however. In the North Tyne area of England, where the NOFU was established early in the war, some complained that local men were "on the dole" while NOFU men continued to work (Home Bred, 1940).

Some experienced lumberers considered the whole enterprise of importing migrant workers "an enormous and quite inexcusable waste of money" (Pitwood, 1940, p. 9). Resistance to the NOFU members also arose based on observations about the Newfoundland Forestry Companies during the First World War. One writer argued that pitwood cutting was not an advanced skill and local unemployed workers "could have been trained in a very short time to equal if not to exceed the output of the Newfoundlanders that were in this country in 1914–1918" (Timber, 1940, p. 9). Some writers to local UK newspapers cited the lack of expertise, inefficiencies, and wastefulness of the First World War unit (Pitwood, 1940; Timber, 1940) as a warning and encouragement to avoid repeating the mistakes of one war in another. These public observations about the Newfoundland Forestry Companies ran contrary to official statements of that unit's productivity and value during the Great War ("Further Appreciation," 1919).

Certainly, local newspaper accounts featuring the NOFU emphasized their expertise, referring to "expert loggers" ("Men Proud," 1940) and "skilled loggers from Newfoundland" ("Newfoundland Loggers' Badge," 1940). Some journalists painted a rosy picture as part of a concerted effort to quell rising misconceptions of the NOFU and to plug the drain of workers from the unit who were unhappy about how they were publicly represented and received, among other things. Nevertheless, in official correspondence related to additional recruiting for the second group of a thousand workers in early 1940, the Dominions Office emphasized the need for men with adequate skills, indicating "some tightening up of selection" was needed (Eden, 1940, p. 1).

Because of these selection issues, Joseph Curran was sent back to Newfoundland in 1940 to oversee recruitment of the second group of men, and to ensure selection met the standards of the Forestry Commission. Curran raised the minimum age for recruitment to 24 years (Ewbank, 1940b), a requirement in line with his observation that "our best men have been men between 26 to 35 years who have had some logging experience and are sound, sensible men. A lot of young men are not desirable. They are hard to control and have lack of patience, which is greatly needed in the work now being done by NOFU" (J. Curran, 1940, p. 3). This move was a way to manage discipline in the NOFU workforce—block those considered most likely to be non-compliant.

The perception of the NOFU had worsened by the spring of 1940, just before the first six-month contract came to an end. At this point, the unit found itself in the crosshairs of the UK Comptroller and Auditor-General, whose report focussed on the expenses paid against the £500,000,000 advanced to the UK government war chest at the onset of conflict. The Comptroller noted the conditions of the six-month NOFU contract: the basic rate, the all-inclusive board, bedding, etc., and the free transportation home. He questioned why men who did not renew their contracts were paid in full while they awaited repatriation but were doing no work. His comments echoed those of Sir Roy Robinson, who considered it "disgraceful behaviour," which the Forestry Commission was powerless to address given the terms of the NOFU contract (Robinson, 1940d, p. 1). It was a rare point of agreement with Turner, too, who was frustrated by advice from the Dominions Office that indicated he had no legal recourse in relation to such workers (Turner, 1940f).

The Comptroller also commented that the NOFU rate of timber production "was low when measured by British standards, despite the higher cost of wages and keep" ("Low Forestry Production," 1941, p. 4). He added: "I inquired as to the progress of experimental measures designed to secure better supervision and closer control of output, and also whether later contracts require payment of full wages to men who refuse to work, but I have so far received no reply." Robinson believed, in fact, that the NOFU men "have cost us a good deal per head to get here and so far we have had nothing, or practically nothing, in return" (Robinson, 1940a, p. 1).

Reported in UK newspapers under headlines such as "Paid for Doing No Work" (1941) and subheadings that read "Foresters Refused To Work—Got Full Pay," ("Abuse of Billet Allowances," 1941, p. 2), the Comptroller's report did little to garner support for the unit, and none of its claims were countered or contextualized by official responses. As the Comptroller noted, the NOFU contract stated that the contract period—pay and service—ran from the day the worker was instructed to leave Newfoundland until the day he returned. Subsequent contracts were revised to provide stronger and clearer language and would stipulate that, at the end of a contract, members would have to continue to work while awaiting transport home. A caveat was also added about such transportation: it would be provided at the earliest possible date. The clause pointed to the previous slow speed of repatriation as having caused some men to refuse to work.

Given the shortages of pitwood and skilled labour, delays in transport home may have felt like a coercive tactic to some members and it stoked suspicions that Timber Supply was trying to squeeze more work out of men whose contracts had ended, or somehow slyly renew contracts against their will (Turner, 1940g). This distrust was aggravated by information distributed by people Sir Roy Robinson called "labour agitators," who visited base camps and encouraged men to refuse to work beyond a forty-four-hour work week—or eight hours a day instead of ten—in line with limitations set for the previous occupants of the base camps (Dominions Office, 1940, p. 1).

Eventually, men who refused to work while awaiting repatriation or were in breach of contract were housed at Bailliemeanoch (Turner, 1940f)—a practice supported by the Chair of the Forestry Commission (Robinson, 1940f) and intended to ensure that these men could not poison an operational camp's atmosphere, increase agitation, or compromise morale. As Turner explained, when there were large numbers of men awaiting repatriation—there were as many as 300 in July 1940 and another 300 anticipated at the second round of contract expirations later that summer (Turner, 1940g)—an atmosphere of disorganization descended on the operations and contributed to decreased production (Turner, 1940g, 1940h), as well as "a rather general feeling of distrust" (Turner, 1940g, p. 1) among the men of the unit.

Robinson's dismissal of worker rights was an example of the critical attitude toward NOFU workers and leadership, which was evident throughout the time the unit was overseen by the Forestry Commission—until the end of 1940. The same attitude, however, was still evident in 1942, when matters related to the unit and its treatment by both the Forestry Commission and the Ministry of Supply came to a head. In early 1942, however, Clement Attlee, a Labour Member of Parliament, became the Secretary of State for Dominion Affairs. In him, finally, the NOFU gained an ally who would recognize their place in the delicate balance of relations with Newfoundland.

Campaign to Remain: Unit Discontent Rises as Official Pressure Builds

As the expiration date of the first six-month contract drew nearer and the war escalated, the UK's supply options for imported timber decreased. By the spring of 1940, trade with Sweden was cut off. Despite this situation, it was clear to NOFU leaders that, for a variety of reasons, they would soon suffer a significant loss of workers. Just when an intensification of local timber production was needed, worker availability was threatened. And so began a desperate unofficial campaign to encourage NOFU men to remain in the unit, which took the form of visits by officials from both the Dominions Office and the NCG.

The Duke of Devonshire, Parliamentary Under-Secretary for Dominion Affairs, was the first to visit NOFU camps in an effort to persuade men to renew their contracts. In April 1940, accompanied by Sir Roy Robinson of Timber Supply, the Duke conducted a three-day tour of camps during which he thanked the men for their work and urged them to stay ("Loggers Urged to Stay," 1940). Publicly he commented on a recurring wartime theme—"the fine types of manhood" the workers represented ("Newfoundland Loggers," 1940, p. 6). Echoing comments J. H. Gorvin had made in his radio recruitment broadcast at the onset of conflict, the Duke emphasized that "in wartime it is up to each man to do the work for which he is best fitted. There are plenty of men for the Army but only these loggers can do this vital work at a time when timber is sorely needed for war purposes and skilled labour is difficult to obtain" ("Newfoundland Loggers," 1940, p. 6).

Sir Roy Robinson had indicated to the Dominions Office that the best way to counter the imminent loss of workers was "to tackle the loggers in bulk by propaganda" (Dominions Office, 1940, p. 1). He saw the problem as "to a large extent psychological" (Dominions Office, 1940, p. 2). In his capacity as Secretary of State for Dominion Affairs, Viscount Caldecote made his presence known to the men in June 1940, by way of a written appeal. The appeal reflected Turner's advice that "print is plain and large and the words not too long" (Ewbank, 1940b, p. 2) and was posted in all the NOFU camps. It included a plea to those considering leaving the unit "not to do so but to stand fast," for "you cannot serve the Crown or your country any better than by remaining at your present work" (Viscount Caldecote, 1940b, p.1).

While visiting camps in late May and early June, Turner issued his own pointed appeal, a no-nonsense and candid description of the dire circumstances in which the UK found itself at the end of the first year of the war. France had fallen, and re-supplying the British Expeditionary Force, which German troops had forced from the Continent, was crucial. About the supply of pit props, Turner explained:

> We must produce more coal at once than we ever dreamed of and—in order to get this coal—must produce not only more pit props than were ever cut in this country, but more than were ever imported. Further, in view of the possibility of railing and road transport being interrupted by bombing, we must—in the next little while—build up reserves of pit props at all mines so that, if the supply is cut off, the mines can keep on producing while repairs are being made and new timber supplies brought in. We must have timber if we are to keep fighting. (Turner, 1940d, p. 1)

Turner called the situation so dire that "only a matter of life or death" was sufficient reason to request repatriation, adding, in an appeal to the men's national pride, that "you can all realize what it will look like if a large number of Newfoundlanders throw up their job and ask to go home when the Empire is in a tight corner and men from all the other Dominions are rushing to take their share in our common cause" (Turner, 1940d, p. 2).

At the time of the Turner and Caldecote appeals, the NOFU was at its peak strength (over two thousand workers); officials in both countries hoped to maintain (or even increase) that number (J. Curran, 1940; Gorvin, 1940c). No other dominion forestry units were yet operating in the UK.

Despite all appeals, however, the NOFU number fell dramatically in the following months. By the fall of 1940, only 1,176 men of the original total (2,151), or 55 percent, remained in the UK. When one thousand new workers arrived in the summer of 1940, the total number of men in the NOFU at the end of its first year overseas was 2,171. This number would decrease steadily over the span of the war.

The NCG Survey

A year after the formation of the unit, in late November 1940, the Commissioner of Natural Resources sent a memorandum to his commissioner colleagues about an informal survey of men who had left the unit. Officials had questioned 390 of the workers repatriated during that year about why they had left the NOFU. Secretary Claude Fraser compiled the data and Gorvin presented a summary to the Commissioners (Gorvin, 1940c).

Key Reasons for Leaving the NOFU (1940 Survey)

Served six months, as intended	177
Accounts not settled[23]	113
Family / personal reasons	35
Health issues	22
Superiors' inattention to concerns	19
Unsatisfactory camp diet	7
Taking a break prior to enlistment in armed forces	6
Employment secured at home	5
Work was unappealing	1
Unknown	3
Dismissed	2

23 An additional forty-nine men also listed pay issues—such as not receiving pay owed to them or receiving inappropriate charges (for clothing, for example)—as a reason for leaving.

Upon review, Gorvin felt the responses did not fully explain the mass exodus of workers, although they did indicate general dissatisfaction—despite almost half of the men surveyed indicating they had intended to stay only for the length of the first contract when they signed. Gorvin concluded from the responses that there were three key issues: accounting, discipline, and organization. Steps to address poor accounting, largely related to outstanding wages, were taken immediately. The goal was to ensure accounts were settled and workers paid in a timely and accurate manner (Ewbank, 1940c; Gorvin, 1940c, p. 3). The issue of unit discipline that was suggested more by unsavoury press coverage than survey data, however, was less easily addressed. For both Gorvin and Sir Roy Robinson, his counterpart in the UK, these problems were directly (if not solely) related to unit organization and the amount of supervision, to which they would soon turn their attention.

Despite their analysis and attention to some of the matters arising from the survey, however, attrition numbers continued to increase. Commissioner of Finance John H. Penson visited ten camps in Inverness-shire in early March 1941, at which time the unit's fate in the UK was not secure. During his visit, Penson observed operations and conditions and impressed upon the men the need to remain at their work. In his report on the visit, he made clear that, in his estimation, the unit's lack of success rested with the Forestry Commission and its successor, the Home Grown Timber Production Department (Timber Production, henceforth) of the Ministry of Supply.

Penson challenged the deference that both Timber Production and the NOFU Officer-in-Charge displayed when unit members decided to transfer or repatriate at the end of a six-month contract. Penson felt that this choice "is not fought seriously" and that Turner appeared to accept it as "almost inevitable" (Penson, 1941a, p. 1). He suggested that Timber Production's authority and that of the Officer-in-Charge be reasserted, and that the department clarify its policies and procedures for the unit, noting that "it is their responsibility to make the Newfoundland logging unit a success," and adding: "I doubt if that is their attitude at present" (Penson, 1941a, p. 3). Thus the NOFU was caught between the competing interests of the NCG and the Ministry of Supply—even though both shared the same goal: to retain woods workers in the UK.

John H. Penson, Commissioner of Finance in the Newfoundland Commission of Government, with NOFU workers in Inverness-shire, Scotland, 1941.(Courtesy of The Rooms Provincial Archives Division, A 8–16)

When he returned to Newfoundland, Penson addressed the country in an April 30 radio broadcast that focussed almost entirely on the NOFU. His thrust was direct: the number of men returning to Newfoundland from the unit was unfortunate and unacceptable and "may have given an adverse impression in Britain" (Penson, 1941b, p. 8). He called upon women, in particular, to write to their men and encourage them to stay. Penson acknowledged that some men had good reason to return, but that such was not always the case. He added:

> We certainly hope that those who are still on the job will continue there. That is where their duty lies, and their interest, too. I told the men plainly that woods work might be a good deal restricted in the coming season in Newfoundland, and that there could be no question of men coming home and then looking for relief. They won't get it. (Penson, 1941b, p. 8)

English-born Penson's remarks summoned, even if not directly, negative impressions about the character of the men akin to those conveyed by the Amulree Commission report, which had linked moral decline among Newfoundlanders and Labradorians to a dependency on the state that compromised their work ethic (Overton, 2001). They also echoed earlier remarks made by his compatriot Gorvin who, when advising Turner that, among the steps that could be taken to stem the flood of departures of men, he should "tell them also that possibilities for employment are not bright and that dole issues may have to be cut down in view of financial situation" (Gorvin, 1940b).

Penson's remarks not only drew on troubling assumptions about the Newfoundland men and their relationship to work, his observations about future opportunities for employment, like Gorvin's a year earlier, were also inaccurate. By 1942, the woods labour supply in Newfoundland was so dire that not only did the NCG refuse a Dominions Office request for an additional unit of men, but there were calls by some to follow the lead of the UK and, to address the shortage of woods workers, hire "lumberjills"—a colloquialism for women woods workers that was coined in the First World War ("Lumberjills," 1942; "Lumberjills for Newfoundland," 1942). British Columbia, Canada, was among the places that would do so during the war.

Workers who did not renew their contracts were met with harsh public criticism at home, leading some of them to comment on "the darn little respect" shown them upon return (Boys of Newfoundland Overseas Forestry Unit, 1940, p. 4).[24] They were referring, in particular, to a scathing editorial published in *The Evening Telegram* that questioned their patriotism. Reinforcing the mythic hold those who fought in the First World War had over those experiencing the Second World War, the writer had wondered: "[I]f there was an important war job still to be done, might it not be expected that Newfoundlanders would stick to it until it was completed? That was the attitude

24 See also Wilson & Williams (1940).

of the members of the Royal Newfoundland Regiment" ("Newfoundlanders Return Home," 1940, p. 6). The editor further commented:

> At a time when the faces of the men of the entire Empire are turned towards Britain and are hasting there with all possible speed to do their part in defeating Germany, it is surprising—it is more than that, it is disturbing—to learn that some hundreds of Newfoundlanders have returned unwounded from the battle. Unless there was no further part that they could play over there, what is the explanation? ("Newfoundlanders Return Home," 1940, p. 6)

Men who left the unit to enlist in combat service were also sometimes criticized. But regarding those who "used the Unit as a means of getting over to join up on this side" (Turner, 1943a, p. 9), however, the NOFU's Officer-in-Charge was magnanimous:

> [T]hey did not let the Unit down. They worked well and faithfully for the six months they had agreed to serve, but the minute that time was up they left us to take what they thought was a fuller share in the war effort. My best wishes go with all of them.... [and] wherever the ships sail, or the planes fly, or landing parties rush the beaches, there will be found Newfoundland boys who crossed the ocean as members of the Unit. (Turner, 1943a, p. 9)

And so it was. Throughout the war, the growing list of Newfoundland casualties included hundreds whose war service began with the NOFU. A particularly sorrowful time in the overseas camps occurred when word arrived of the sinking of the HMS *Hood* on May 24, 1941. When the German battleship *Bismarck* fired and hit the Royal Navy battlecruiser, the largest in the world, only three of the 1,418 people aboard survived. It was the largest single-day loss in any Royal Navy battle during the war. Sixteen of the crew were from the Island of Newfoundland, twelve of whom had served in the NOFU (Tarrant, 2008).

Increasing Dissatisfaction: Leadership, District Organization, and Unit Discipline

When Commissioner Gorvin shared the survey results of NOFU members leaving when the first six-month contract ended, he wrote extensively of discipline and organization, two issues he felt were at the heart of discontentment within the unit. According to the survey respondents, foremen and sub-foremen were sometimes indifferent to their concerns and prioritized their own entertainment and enjoyment over the needs of the men they supervised, which led not only to widespread dissatisfaction but also low productivity. The foremen, for their part, pointed to poor behaviour among the men. It was a version of the distrustful and antagonistic dynamic between fishers and merchants all too present under the truck system in Newfoundland in the 1930s (and earlier): merchants who acquired the fish by supplying needed goods (rather than cash) accused fishers of dishonesty, and fishers accused merchants of depriving them of legitimate earnings (NRC *Report*, 1933). Gorvin observed that "the lack of discipline appears to have been general" and, he claimed, had been anticipated, given the civilian nature of the unit (Gorvin, 1940c, p. 4)—a decision that had been made by the Dominions Office. Gorvin was frank in his criticism, while also showing a rare defence of the men:

> In the first place the Unit should have been of a military character considering its size. It is rare for any organization to import labour on such a scale on civilian terms, especially when that labour will be operating under completely new circumstances and in entirely different conditions. The men themselves thought of the work as a war effort of six months' duration and felt entitled to return when that period was over. Had they been signed up as part of a military unit, given uniforms, and been required to live up to an exact routine, this story might have been different: it would have been easier to persuade them to stay on. (Gorvin, 1940c, p. 4)

The idea of converting the NOFU to a military unit had been raised several times once the unit was overseas, but it was always considered unfeasible (Gorvin, 1940c; St. John Chadwick, 1942b). As an alternative, Gorvin suggested that wartime labour regulations be applied to the unit. Beginning with the Emergency Powers (Defence) Act of 1939, introduced after the war began, the UK government passed a series of legislation to ensure an adequate labour supply for war industries.[25] Gorvin's point preceded the introduction of one of these policies—the Essential Work Order, on March 5, 1941—but his comment indicated that the NCG was open to having relevant legislation applied to the NOFU to address unit discipline.

Sir Roy Robinson of the Forestry Commission had also flagged inadequate supervision as a problem in the early months of the unit's arrival in the UK. In a June 21, 1940, letter covering Turner's report for the month of May, Robinson drew special attention to the need to increase the supervisory depth of the unit at both the higher level of superintendents and the subordinate level of foremen. Robinson wanted to increase the number and the quality of supervisors so as to improve both unit productivity and discipline. The issue was even more pressing when additional workers were sought for the unit in 1940 and again in 1941—adequate numbers of supervisors, Robinson emphasized, were needed to accompany the new workers (Robinson, 1940c). He expressed his concern about the problem of ineffective and insufficient supervisors strongly to the NCG, noting that one of the first group of superintendents was "particularly weak" (Robinson, 1940e, p. 1). He concluded: "Personally I consider it even more important to replace (and if necessary to send back to Newfoundland) inefficient Superintendents and foremen than to send back unsatisfactory men" (Robinson, 1940e, p. 3).

The supervisory system in place was established upon arrival in the UK: the NOFU had been quickly organized into four districts (T. V. Curran, 1987), each with a district headquarters led by a superintendent. The administrative headquarters was at Edinburgh, where offices of the Forestry Commission and Timber Supply were also located. Until late 1941, both Turner's office and

25 For a discussion of this legislation, see Schoenfeld & Whitney (1942).

that of the NOFU Chief Operational Superintendent were in Edinburgh; they were then relocated to Carrbridge, which was closer to operations.

Original NOFU Districts

District	Headquarters	Superintendent
One	Beauly Later, Muir-of-Ord	Thomas Curran
Two	Carrbridge	Alleyne Hubbard
Three	Ballater	Edgar Baird
Four	Kershopefoot Later, Galashiels	Bruce Fraser Later, Charles Cahill

Following discussions with Timber Supply, the NOFU organizational grid was reconfigured in the summer of 1940 to address issues of supervision and discipline. Three additional districts were created and a new superintendent appointed to each one. They were: Brendan F. Davis of Bishop's Falls, who had been a timekeeper with the ANDCO; Clarence Lingard of Bishop's Falls who had been a roadbuilder with the ANDCO; and, Peter J. Petipas of Whitbourne, who had been a steelworker in New York for thirteen years then farmed at Whitbourne, Newfoundland, in the mid-1930s. Davis, Lingard, and Petipas were gazetted on October 20, 1942 ("Public Notice," 1942), along with Thomas Curran, who had been promoted to superintendent upon his arrival in Scotland in January 1940. As part of the reorganization, Charles Cahill of St. John's, who had joined the Department of Natural Resources as a timber surveyor and cruiser in 1934 and became Chief Forestry Surveyor in 1936, replaced Bruce Fraser at Galashiels.

Additional NOFU Districts (Summer 1940)

District	Headquarters	Superintendent
Five	Kingussie	Peter Petipas
Six	Lochgilphead	Clarence Lingard
Seven	Perth	Brendan Davis

By late spring in 1940, also at the recommendation of Sir Roy Robinson, the two NOFU camps in the New Forest area of England—at Fritham and Brockenhurst—were placed under the technical supervision of the Forestry Commission. Their distance from the NOFU headquarters created supervision challenges, as did the organizational differences between the Forestry Commission of Scotland and that of England. Turner had wanted the 302 men in those camps moved to Scotland, as part of the intensification of the unit's work there, but his recommendation was refused. At that point, four additional men were loaned to the Forestry Commission—in this case to build sawmills—raising the total of such men to 77 (Turner, 1940f).

By late 1942, camps in southern Scotland were closing and NOFU operations were concentrating in the north. These changes led to yet more reorganization of districts (T. V. Curran, 1987). The unit was also responding to changes in other dominion forestry operations, in particular, the British Honduras Forestry Unit and the Canadian Forestry Corps. The former was repatriated in early 1943 (Forbes, 2015),[26] and the Canadian Forestry Corps camp numbers were diminished by a third in the fall of 1943 (when men were returned to Canada) and again with the reassignment of personnel strength to Europe with the Canadian Army prior to D-Day (Wonders, 1987, 1991). The NOFU took over the operation of the camps vacated by the British Honduras Forestry Unit as well as a few vacated or unfinished by the Canadian Forestry Corps.

Final NOFU District Organization (Fall 1942)

District	Headquarters	Superintendent
A	Muir-of-Ord, Ross and Cromarty	Brendan A. Davis
B	Carrbridge, Inverness-shire	Thomas Curran (to 1944)
		James G. Martin
C	Kingussie, Inverness-shire	Peter J. Petipas

26 For an account of the controversial circumstances of this repatriation, see Forbes (2015).

At the end of 1940, the strength of the NOFU was 2,171 men—approximately 1,000 of whom had signed on for the war's duration—777 men had returned home and 200 had transferred to the armed forces ("Fighting Sons," 1940). By the end of 1941, of the 3,595 men who had joined the NOFU since its formation, 529 had transferred to combat forces—most at the end of the first contract before the restrictions to prevent such bleed were implemented—and a total of 1,127 had returned to Newfoundland. For the same period, the strength of the unit was reported to be 1,876 men ("30,000 Men," 1941, p. 7).[27] At the end of 1942, this number had shrunk to 1,673. By the end of 1943, the strength of the unit had decreased to 1,320 and at the end of 1944 numbered only 1,180 (Turner, 1945b).

NOFU Statistics[28]

End of Year	Strength	Repatriation	Transfers	Deaths	Dismissals
1940	2,171	777	200	6	
1941	1,876	350	329	17	42
1942	1,673	173	209		23
1943	1,320			7	
1944	1,180		4	2	
1945				1	
1946	≈1,000			1	

27 Based on statistics compiled from various government sources and reported by *The Daily News*, December 31, 1941, p. 7, and including deaths, dismissals, and men loaned to the Forestery Commission.

28 These statistics are compiled from various sources (T. V. Curran, 1987; "Fighting Sons," 1940; "30,000 Men," 1941; Turner, 1945b). The number documented for overall enrolments varies across sources, but 3,595 is a good proximate figure. Numbers are not available for all categories, i.e. dismissals and repatriations, or for all years. Some numbers were deduced from available figures.

"Far From Satisfactory": Improving Productivity

At the start of 1941, oversight of the NOFU was moved from the Forestry Commission to the Ministry of Supply, under its new Home Grown Timber Production Department—the name of which clearly captured the wartime priority. Its central office was in Bristol, England, and was led by Director Sir James Calder until the end of 1941 and his successor, Gerald Lenanton, thereafter. The Scottish branch of Timber Production was headed by a Deputy Director, Sir Samuel Strang Steel, with General Sir Harold Carrington next in command. Like Sir Roy Robinson of Timber Supply, both Calder and Lenanton were critical of the NOFU, regularly pointing out low productivity, inefficiencies, and ineffective leadership.

Productivity had been a contentious issue from the moment the unit began its work. Along with the public's questions about production quantity and work quality (Pitwood, 1940; Timber, 1940), and criticism from the Auditor-General ("Low Forestry Production," 1941), there had been a steady stream of correspondence about these things between the Dominions Office and the NCG, as well as between the Dominions Office and Timber Supply. "The unifying point was that, by UK standards, production was low and inadequate to meet the timber supply crisis at hand" (Robinson, 1940d, p. 1). Officer-in-Charge Turner attempted to counter what he felt were comments that left false impressions of his unit and his leadership. In response to Robinson's cover letter to his October 1940 report, for example, Turner wrote that it was likely to leave the NCG with "the impression that the unit is a patch of inefficiency in an otherwise perfect organization, and that I am entirely responsible for this unhappy state of affairs" (Turner, 1940j, p. 1).

In a lengthy memorandum to Governor Walwyn, written prior to the recruitment of the last group of four hundred workers in the spring of 1941, Secretary of State for Dominion Affairs Cranborne outlined several concerns related to the NOFU contract as well as other items arising from the Timber Supply's and the Ministry of Supply's experiences with the unit. Cranborne hoped the concerns would be addressed when the next contract was developed and during the recruitment of the next group of workers. The issues ranged from the amount of the allotments and income tax payable to the contract's terms, the recovery of costs for transporting home men who were in breach of their contract, and fines or pay cuts for below-standard work. Cranborne noted:

> You should know in this connection that Ministry of Supply are not at present very satisfied with production of unit and they attribute this to some extent to recruitment amongst early contingents of a number of undesirables. They attach great importance to careful selection from amongst men whose records are known to be satisfactory and who have had some experience at logging. (Viscount Cranborne, 1941a, p. 3)

After returning from a visit to Scotland in March of 1941, NCG Commissioner of Finance Penson also drew attention to productivity issues in an address to the people of Newfoundland. Describing production in the camps as "far from satisfactory" for several reasons—some "quite outside the control of the men"—Penson supported the introduction of payment by piece rates rather than by time, to provide incentive for increased efficiency; as Turner earlier reported to the NCG (1940i), he announced that this measure was already being adopted in some instances (Penson, 1941b, p. 8).

Penson's focus on piece work pay rates to the exclusion of any mention of other specific issues suggested that worker performance was the major contributor to productivity problems. It was a familiar refrain: the men were of questionable character and quality, undisciplined and psychologically weak, and loathe to work unless externally motivated. But the situation was far more complex than pay structure. There were many issues affecting productivity that were, as he noted, out of the control of the unit, a point on which Penson had chosen not to elaborate. They included weather, forest stand quality, debris removal policies, equipment shortages, equipment maintenance, shipping delays, and member and leadership changes (Turner, 1945b).

Adverse Weather

The winter of 1940 was not the only time when weather greatly inhibited timber production, but for the NOFU it coincided with their orientation to woods work in a new country and so presented particular challenges. January 1940 was the coldest month that Scotland had experienced in forty-five years. Heavy snowfall also impeded the unit's efforts to get established, as this weather summary for the year indicates:

> Snowfalls in the last week of January and the first five days of February were exceptionally heavy, trains being "lost" in deep drifts and load traffic virtually at a standstill. Hundreds of small places were cut off for days and in at least one instance food was dropped by parachute to isolated troops. Although falls of snow were not heavy after the first five days of February, nevertheless it did fall no less than 22 days in the month. Again in March snow was frequent and widespread ... ("Weather in 1940," 1941, p. 3)

The harsh winter weather in 1940 prevented the movement of men and supplies (Turner 1940a) and made for a sluggish, inefficient start to the work of the unit on which the UK was depending—the NOFU was the only wood-cutting unit at work in the country at the time. The small number of NOFU men who remained farther south, in England, fared better weather-wise, reporting less snow, but the damp climate and poor heating were considered factors in the high number of respiratory illnesses they suffered that winter (J. Curran, 1940; "Letter from a Local Boy," 1940; Turner, 1943a).

Nor did things improve greatly in the following year.[29] As one correspondent wrote about the second group of the NOFU, which arrived in summer 1940, they also had "a somewhat harsh introduction" to the Scottish

29 Tom Hall, NOFU area office manager at Ballater and author of several books of walking tours, penned a description of a trip between Ballater and the two Kildrummy camps made during the winter of 1941 to pay the men working there. It captured his incredulity at snow conditions that made travel so perilous and difficult and, in this case, turned what was typically a one-day return trip into a three-day affair. See Hall (1941).

climate—according to locals, it was among the hardest winters known in Scotland in the century. Despite being familiar with extreme cold and high snow accumulation, the NOFU men were less accustomed to "the raw air" that swept the Highlands ("Lumberjacks: The Newfoundlanders," 1941, p. 7). The late fall and early winter of 1944 were also "exceptionally bad" (Turner, 1945b, p. 7).

Weather was also an inhibiting factor for other units. The camps that had been worked by the British Honduras Forestry Unit until 1943, which were then used by the NOFU, were in west Highland areas of high rainfall. Along with other contributing factors there, it meant lower productivity at these camps, which in turn negatively affected overall unit productivity rates (Turner, 1945b). The year-round weather was considered "the most serious problem" for the Canadian Forestry Corps, which arrived in 1941 (Wonders, 1987, p. 28). Their accounts have extensive descriptions of poor conditions: excessive rain, washed out roads, muddy work sites, snowstorms, and heavy snow accumulations (Wonders, 1987, 1991).

Timber Stands and Forest Policies

The forest stands initially identified for felling by the NOFU were often sparse, resulting in low yields. It was not until the Forestry Commission secured large stands, mostly in the north of Scotland, that this issue was resolved. These stands included Scots pine, spruce, larch, and fir, along with some hardwoods.

To harvest the trees, the unit had to adhere to existing forest-use policies. They included the requirement to cut to the forest floor, leaving no stumpage, which reduced waste but increased strain on workers. In an age before power tools, the men used crosscut saws or, most often, a bucksaw—what some of the men called the "man-killer" ("Tough Men," 1941, p. 2). The NOFU introduced bucksaws to Scotland, where cutting had traditionally been done by axe. Initially, the men were also required to remove "slash," the debris or by-products of cutting. Once this labour-intensive requirement was lifted, workers were freed for other tasks that improved unit productivity (T. V. Curran, 1987).

During an inquiry into conditions in the NOFU camps in 1942, Joseph Curran expressed the frustration within the unit about productivity issues

Two NOFU workers cutting and felling with a crosscut saw, 1940, Scotland. (Courtesy of The Rooms Provincial Archives Division, A 8–13)

and their impact on morale and reputation to Charles Peat, Parliamentary Secretary to the Minister of Supply. He drew attention to the nature of the timber stands the unit had been assigned—in difficult places that other units had refused. The unit shared this grievance with the British Honduras Forestry Unit, when they inherited their camps in 1943. Curran also called out government officials' deference to estate owners whose timber stands had not been fully requisitioned and, thus, caused NOFU camps to be relocated despite timber availability where they were. He also noted the time and labour wasted by requiring the men to burn slash, adding that it was also inappropriate to expect the NOFU to repair estate walls and deer fences. Solid reasons for these questionable practices were not communicated to the unit, thus creating another fault line in early relationships between it and the UK officials who oversaw timber production (Peat, 1942b).

Access to Equipment

The equipment available to the NOFU in the early years of the war was inadequate in quantity and kind. Without ample Caterpillar tractors and trucks, the unit had to make do with the low-impact machines traditionally used by the Scots, such as the rubber-tired Fordson tractor, as well as horses. When the Fordson was found to be impractical on wet, slippery, and hilly terrain, other tried-and-true Newfoundland ways were found for moving timber. Chutes[30] and tramways, short temporary rail lines or sidings over which small four-wheeled rail carriers (trams) operated, were among them.

Horses—often garron ponies—were a mainstay for skidding or hauling felled wood short distances. The "go-devil," a small horse-drawn sled once in common use in Newfoundland, was also a novelty to the Scottish people. It was constructed of two runners joined by a crossbar onto which the butt ends of logs were secured; they could be thus easily hauled, especially over snow. Early in their work, the NOFU men had to convince the Forestry Commission of the viability of the go-devil to compensate for shortage or inadequacy of equipment (Turner, 1940e). On some of the soggy ground in the New Forest, one NOFU worker suggested to a reporter that, instead of the proposed tractor, "if we could harness up some of those ponies around here, we could build a sled in a couple of hours and get all the stuff out without any trouble"—which they did ("Over the Teacups," 1940, p. 4).

A NOFU worker hauling logs with a garron pony, 1940. (Courtesy of The Rooms Provincial Archives Division, A 7–169)

30 The Newfoundland Forestry Companies had introduced chutes in their work overseas during the First World War. One of its chutes—approximately 3,200 feet (or 975 metres) in length—was considered at the time to be the longest ever made.

NOFU workers loading pit props onto a tram, 1940. (Courtesy of The Rooms Provincial Archives Division, A 8–14)

By 1941, machinery—including heavy-duty tractors, trucks, and bulldozers—became available in greater numbers and, by 1942, when the NOFU equipment branch was integrated into the Ministry of Supply, there was adequate equipment to maintain and increase production levels. By end of 1942, the unit had also constructed a central repair depot staffed by thirty workers who lived on-site (Turner, 1943a). The arrival of the Canadian Forestry Corps also enhanced access to equipment. In response to particular shortages in each unit, the two units entered into an agreement whereby the NOFU delivered raw timber to mills run by the Canadian Forestry Corps (which was suffering a supply shortage), in exchange for the use of its equipment for harvesting, transporting, and roadbuilding (T. V. Curran, 1987), thereby improving efficiencies for both units (Turner, 1945b). This collaboration marked an improvement in relations from 1940, when Turner had warned the Dominions Office that plans to locate camps for the Canadian Forestry Corps among the NOFU locations would upset his men. He advised that they were best kept in separate districts (Ewbank, 1940b). The agreement between the two had the unfortunate side effect, however, of reproducing a hierarchy of workers with national overtones.

Transport and Shipping

Efficient ways to transport logs and ship milled lumber were essential components in maintaining unit productivity. Harvesting sites had to be accessed via roads, which had to be built along with the camps themselves. Harvested timber was moved from the cutting site to a roadside, and from there by lorries to a mill or a shipping site, usually a train station or a vessel port—neither of which was dependably ready to receive it. Transportation equipment required upkeep and repair, so demand for carpenters, mechanics, and parts was high (Turner, 1945b). The Canadian Forestry Corps experienced similar problems, termed as "the other major complication" of their time overseas (Wonders, 1987, p. 28), and encountered similar challenges—a lack of a road network for transport, damage to and upkeep of new roads, inadequate rail-loading facilities, and disrupted shipping schedules.

By the end of its work in the UK, the NOFU had developed an extensive network of roads that covered over 200 miles (322 kilometres) and provided access to timber stands and camps. The "Panzer Division," as the road crew was named—likely because a convoy of lorries resembled a German tank brigade by that name—built a road to a camp under construction and then built access roads for harvesting and extraction as needed. The road crew was located at a central depot camp for easy access to operational camp sites as needed (*Operations*, n.d.). Turner noted in his 1945 unit review that "some of the roads are now over two miles long, tapping some difficult territory, and a good deal of labour was used in this construction and in maintenance of these roads" (Turner, 1945b, p. 5). Road construction and maintenance, like other aspects of the unit's work, were greatly affected by weather; inclement conditions often delayed the building of new roads or damaged existing ones.

At the more northerly and very isolated NOFU camps near Ullapool, which had been inherited from the British Honduras Forestry Unit, transport and shipping problems were greatly exacerbated. Timber extracted at these locations was transported by coastal steamer rather than rail, and the distance to be covered between cutting sites and shipping ports, along with often poor conditions, reduced transport efficiency from these locations

(Turner, 1945b). Despite the low production, high labour demands, and the general inefficiencies of these camps, so desperate was the wartime need that they nevertheless were important sources of timber supply. With its newly received responsibility to harvest these difficult sites—known to be a liability for productivity—the NOFU was still able to maintain overall impressive numbers in the second half of its time overseas.

Despite the obstacles and frustrations of the NOFU's first years in the UK, by early 1943 the Department of Natural Resources was able to release an optimistic report on the unit, which was written by Officer-in-Charge Turner. First published in local newspapers, his report was part of a public relations effort in both countries intended to quell some of the disquiet about the treatment of the men and their reputation, and to create a positive image of the unit and its accomplishments (St. John Chadwick, 1942c). Turner called attention to the fact that the unit had "earned the distinction of being the most efficient now operating in the United Kingdom and its output per man employed is higher than that of any similar unit" (Turner, 1943a, p. 9). By the end of the war, the unit had harvested one million cords of timber and produced 70,000,000 cubic feet (or 1,982,179 cubic metres) of lumber (T. V. Curran, 1987).

During the period covered by Turner's report, however, the unit was still experiencing a significant loss of workers; its strength was diminishing by the hundreds every year since its formation. The Ministry of Supply was so concerned about attrition that it considered invoking the newly minted Essential Work (General Provisions) Order, which required workers considered essential to have the National Service Officer's approval to leave their employment (Viscount Cranborne, 1941c).

It was not only the regular members of the NOFU who were making decisions to leave. There were several leadership changes in the first years of the unit. All but one of the initially assigned superintendents had left the unit by 1943. Bruce Fraser took a position with Bowater-Lloyd, the company for which he had worked in Newfoundland. Edgar Baird transferred to the Royal Air Force in the spring of 1941. In the summer of 1942, Alleyne Hubbard transferred to the Canadian Forestry Corps and, two years later, was killed by shrapnel from a roadside bomb in Nijmegen, Netherlands. And in the summer of 1943, Joseph Curran returned to Newfoundland. Only Thomas Curran remained with the unit for the duration of its work in the UK.

Comrades in Cutting: Other Forestry Units

The NOFU was neither the only nor the largest unit to contribute to timber supply during the Second World War. Units from other dominions also worked in the UK, as well as on the Continent. They included the Australian Forestry Group UK, the British Honduras Forestry Unit, the Canadian Forestry Corps, and the New Zealand Forestry Companies.[31] Alongside them were the British Forestry Companies, who worked both at home and on the Continent, and the Women's Timber Corps of the UK. In addition, workers—prisoners of war, students, and Irish labourers among them—were also involved in timber supply (Vickers, 2011).

The Australian Forestry Group UK

At the onset of war, the UK government asked Australia for three forestry companies of two hundred men each to support the British Expeditionary Force in France. These companies were uniformed and trained as part of the Royal Australian Engineers of the Second Australian Imperial Force. Two companies were raised in 1940 and arrived in England in July of that year. Following the fall of France and the Battle of Britain (July through September 1940), they were deployed instead to Northumberland, England. A third company was raised in the spring of 1941 and it arrived in the UK that August. All three companies were sent to Dumfriesshire, Scotland, and were reassigned to Sussex, England, shortly before repatriating early in the fall of 1943.

31 As with the NOFU, the stories of these units are not widely documented, resulting in gaps and, therefore, inconsistencies in the information available for each unit.

During its UK assignment, the Australian Forestry Group produced 30 million board feet (70,792 cubic metres) of lumber for the war effort (G. M. Long, 1952; McKenzie-Smith, 2018).

The British Forestry Companies

Most of the foresters of the UK were unavailable for recruitment into the British Forestry Companies, which was established in 1939 to supply timber to the British Expeditionary Force. At the start of war, foresters were listed as a reserved occupation and few could be spared to assist the armed service. Most who were recruited were drawn from Scottish estates and the Royal Scottish Forestry Society. By late 1939, the 129th British Forestry Company, attached to and wearing the uniform of the Royal Engineers, was formed; it was deployed to France in December 1939. Two more companies—the 130th and 131st—were raised in 1940, but they could not be deployed on the Continent following the fall of France. The 129th British Forestry Company retreated from France in June 1940. All three British companies were posted in England until 1944, when two were redeployed to Normandy (Lloyd, 1944, 1945).

The British Honduras Forestry Unit

This civilian unit from the country now known as Belize arrived in the UK in two contingents, the first of approximately five hundred men early in the fall of 1941 and the second of approximately three hundred late in the fall of 1942.[32] Like Newfoundland and other dominions, British Honduras had suffered massive unemployment during the Great Depression when its key mahogany export market collapsed. The British Honduras Forestry Unit offered men an opportunity for employment as well as service, and it was quickly seized upon by the men of the country (Forbes, 2015). The unit uniform—dark blue serge battle suits and black berets—was provided by the Ministry of Supply upon the unit's arrival in the UK (Fitzgerald, 1942a).

Many of the experiences of the British Honduras Forestry Unit in Scotland were framed by racial constructs and racism (Forbes, 2015). Newspapers

32 For a member account of the unit, see Ford (1985); see also an interview with member Sam Martinez ("The Last Lumberjack," 2006).

carried stories of raids on the camps during which local women visitors were removed and charged with trespassing ("Third Raid," 1943). Accommodations in Edinburgh were elusive for unit members, prompting the government to build British Honduras House, a hostel for the men during their visits ("Colonial Foresters," 1943). After only two years of work in the UK, the British Honduras Forestry Unit was disbanded (September 1943) because of concerns about venereal disease and miscegenation, an indication of the extent to which racist attitudes and behaviours shaped decision-making about the unit (Bilgrami, 2004; Forbes, 2015). The NOFU began operating its camps in early 1944 (Turner, 1944).

The Canadian Forestry Corps

In the First World War, Canada had mounted a significant response to the UK request for woods workers with the formation of the Canadian Forestry Corps, an estimated 33,000-strong unit (woods workers plus attached personnel and labourers) that provided as much as 70 percent of the Allied wartime timber supply (Bird & Davies, 1919). As a response to the appeal sent in the fall of 1939, the Canadian government established a new Corps in 1940 as part of the First Canadian Army to support the Allied cause in the Second World War. Thirty uniformed and combat-ready companies of the Canadian Forestry Corps were sent to Scotland, with an advance company arriving in the fall of 1940. By spring 1941, there were thirteen companies in place; seven followed that summer (Wonders, 1987, 1991).

The Canadian Forestry Corps was a modern, well-equipped corps; its men were concentrated in the eastern Scottish Highlands. Self-named the "sawdust fusiliers" because of the hills of sawdust that accumulated from the use of their diesel-powered saws, the Corps' headquarters was initially located at Blair Atholl and then moved to Beauly in the winter of 1941. Its numbers peaked at 6,991 in early 1943, after which several companies were returned to Canada to address woods labour shortages there ("Forestry Troops Will Return," 1943); by early 1944, other companies were redeployed in preparation for D-Day. The ten companies that remained worked in Scotland until the end of the war, and all personnel had returned to Canada by the summer of 1945. The Canadian Forestry Corps experienced many of the obstacles and frustrations noted by the NOFU, but none of the aggravations the Newfoundland unit experienced with Timber Production. During its service it produced 2,208,613 cubic metres (77,996,432 cubic feet) of lumber (Wonders, 1987, 1991). The unit was disbanded on December 3, 1945.[33]

The New Zealand Forestry Companies

New Zealand responded to the UK's request for woods workers by forming three forestry companies who wore the uniform of the New Zealand Corps of Engineers—the 11th, 14th, and 15th companies. Together they constituted a war effort of 645 men, all of whom received basic military training before deployment. The first company arrived in the UK in June 1940. Following a short stint of tree-cutting in preparation for the building of an anti-tank ditch in the south of England, the 11th Company was sent to Gloucestershire, where its headquarters were established, and Wiltshire.[34] The two additional companies arrived in November and were also assigned to the south of England—Gloucestershire, Wiltshire, and Hampshire. During their short service in the UK, the productivity of the New Zealand Forestry Companies was affected by cutting and debris disposal policies, much like the NOFU experienced. In the summer of 1943, the three companies were reorganized for service in Algeria and Italy or repatriation to New Zealand by fall 1943. The unit produced 92,180 cubic metres (3,255,306 cubic feet) of lumber during its service in the UK (McKelvey, 2001).

The Women's Timber Corps

The Women's Timber Corps had a precursor in the Women's Forestry Service of the First World War. Approximately two thousand women had participated in that service, working alongside the Canadian Forestry Corps and the Newfoundland Forestry Companies. The Women's Forestry Service was

---

33 For an account of the unit based on the letters home of member Pat Hennessy, see Jarratt (2016).

34 Anti-tank ditches or trenches were part of the line-of-defence preparations for a German invasion anticipated after the defeat of the British Expeditionary Force in 1940.

formed in 1917 as a division of the Women's Land Army, which participated in farming, foraging, and forestry (C. Scott, 2017). Notwithstanding the success of this unit, the UK Government was slow to instate a separate forestry service for the Second World War, despite the dire shortage of forestry sector labour.

At the beginning of the Second World War, the so-called "lumberjills" did forestry work as part of the scope of duties assigned to the Women's Land Army. As the devastating warfare in the Atlantic wore on and the timber supply crisis did not subside, the Ministry of Supply formed the Women's Timber Corps as a separate entity in April 1942 for work in England, Wales, and Scotland (Vickers, 2011). The Corps' members wore the green and khaki of the Land Army, with a distinctive green beret and badge—an oval surrounding a tree and topped by a crown. Women cut, felled, snedded (trimmed), hauled, processed, and scaled (measured) timber. They operated heavy machinery, including tractors and trucks, and worked in lumber mills. Estimates of the numbers who served vary from a low of six thousand (Vickers, 2011) to a high of fifteen thousand (Foat, 2019). The unit was disbanded in 1946.[35]

During the war, the work of the Women's Timber Corps was noticed abroad. Facing a large-scale shortage of loggers to provide wood for pulp and paper production at Grand Falls and Corner Brook because of the drain on the sector created by American base construction, some in Newfoundland floated the suggestion of hiring women to do logging work. Editorials in *The Western Star* in 1942 suggested that women take the work places of the men who had gone overseas in the NOFU, and even suggested that their uniform be modelled on that of the Women's Timber Corps. Noting the eagerness of women to contribute to the war effort, the editor pointed out that "our lumberjacks are overseas now, fighting our fight, and it is our duty to keep the wheels of industry turning while they fight. Newfoundland girls, given the opportunity, will be proud to don uniforms and swing their axes for Victory" ("Lumberjills," 1942, p. 2).

But the NCG, and particularly Commissioner of Natural Resources P. D. H. Dunn, disapproved of women being employed as woods workers. In 1941,

35 For accounts of work in the Women's Timber Corps, see Foat (2019), who interviewed sixty former members as part of the research for her book *Lumberjills: Britain's Forgotten Army*.

he rejected a proposal by Bowater to this effect (High, 2003). Other dominions demonstrated more openness to women in woods work. Lumberjills were hired in the forests of British Columbia, Canada, and women of New Zealand also worked in the sawmills of Bruce Bay during the war ("Lumberjills Beat," 1942).

Forestry Units Operating in the UK During the Second World War

Unit Name	Type of Unit	Approximate Unit Size	Locations	Years
Australian Forestry Group UK	Military	550	Scotland (Dumfriesshire) England (Sussex, Northumberland)	1940–43
British Forestry Companies	Military	≈450	England (Gloucestershire, Nottinghamshire)	1940–44
British Honduras Forestry Unit	Civilian	800	Scotland (East Lothian, Berwickshire, Dumfriesshire, Galloway, Sutherland, Wester Ross)	1941–43
Canadian Forestry Corps	Military	6,990	Scotland France	1941–45
New Zealand Forestry Companies	Military	645	England (Gloucestershire, Wiltshire, Hampshire)	1940–43
Newfoundland Overseas Forestry Unit	Civilian	3,595	England (Northumberland) Scotland (Inverness-shire, Aberdeenshire, Ross and Cromarty)	1939–46
Women's Timber Corps	Civilian	6,000–15,000[36]	England Scotland Wales	1942–46

36 This figure represents the most conservative estimate (Vickers, 2011); others are as high as two-and-one-half times this number (Foat, 2019).

1942: A Crisis in Welfare and a Threat to Imperial Relations

Describing the NOFU in the context of the large numbers who had signed up for wartime work service, Jack Murphy, Assistant Trade Commissioner for Newfoundland in London, commented: "The Newfoundlanders came over in cities, and none of them have been more welcomed than the lumberjacks" (Provincial Archives, 1986, 21:18). Such aplomb made for effective propaganda, but it did not accurately convey the experiences of the NOFU who, in the first years of the war in particular, felt neither welcomed, valued, nor well-treated by the UK authorities who oversaw their stay. And despite being first on the ground when the war began, they were among the last to be recognized for their efforts after it ended.

Even a cursory review of the work of the various wartime forestry units that operated in the UK reveals differential treatment. It occurred at various levels, including the location and type of work assigned, camp conditions and amenities, reception by the locals, and relations with UK governing departments. The NOFU and the British Honduras Forestry Unit stand out as colonial units that experienced the greatest difficulties in all these regards.[37]

Based on some newspaper reports about the NOFU camps, a reader might easily be convinced that workers were alight with glee as they felled timber to supply the Allied war effort. The cultural features of their work at home had become hallmarks of their coverage in the UK press. But stories that focussed on the men's unique language, tools, food, and garb—which usually ended with a description of a colonial evening in the backwoods—belied a much more difficult experience, one in which the NOFU men fought for improved welfare conditions and unit recognition. As the Ministry of Supply focussed on crises of retention and production, the Dominions Office was fielding a different and growing crisis, as men abandoned the unit because of the conditions they endured and the treatment they received from government officials, especially those who worked in Timber Supply and, later, Timber Production.

37 See Forbes (2015) for a short comparison of the experiences of these two units while in the UK.

For a number of reasons, 1942 was a pivotal year for the NOFU. By then, many of its members had worked in the UK for two years without leave, and in extremely isolated conditions. That winter, the esprit de corps in the NOFU camps was at an all-time low. The men were dissatisfied with the lack of amenities, including recreation facilities and canteens, and the limited access to nearby villages for religious or leisure purposes. And while many locals were both welcoming and appreciative of the role the NOFU was playing in the larger wartime effort, others criticized their non-combat, un-uniformed status. Furthermore, the camps had become a public health concern, drawing yet more unwanted and unsavoury attention to the unit.[38]

Sanitation in the NOFU camps in Inverness-shire became a public health issue as early as the spring of 1941. At that time, Sanitary Officer William MacDonald brought the problems to the attention of NOFU officials in Edinburgh; notices were subsequently sent to camp foremen to address the issues. Dr. A. M. Fraser, Chief Medical Office of Health for Inverness-shire, reported that while some camps were better than others, the general level of sanitation was "exceedingly low" ("Forestry Camps—Question of Sanitation," 1941, p. 3). NOFU Administrative Officer Alfred Hart followed up on concerns about camp sanitation. For example, in September 1941, Hart visited Rosebery camp, Midlothian. He described conditions there as "filthy" and "disgraceful" and the foreman as "disgruntled" and "lacking authority in the camp" (A. E. Hart, 1941b, p. 1). By March 1942, following two additional visits by Hart, conditions had improved at Rosebery (M. J. Taylor, 1942), underscoring how increased supervision was necessary to ensure member well-being, as well as productivity. With camps now clearly on the radar of health officials and the public, conditions were more closely monitored and improved overall.

There were also serious gaps in services provided to the unit that were automatically available to the armed forces. Access to canteens, for example, was not available to the NOFU because of its civilian status. The well-being of the men was a concern among community groups in Scotland, including

---

38 The main file that includes reports on the investigation into welfare conditions of the British Honduras Forestry Unit and the NOFU—AVIA 22/1239, which was housed in The National Archives of the UK—was listed as missing.

the Church of Scotland; its members were aware that most of the NOFU camps lacked adequate supports for the workers' social and spiritual needs. Considering that the public and the NCG were attuned to poor camp conditions and timber and labour needs in the UK were at desperate levels—with requests to the NCG for sawyers and a second unit going unfulfilled—and that delicate relations between the imperial government and one of its colonies were part of the mix, the question of the NOFU and its discontents grew ever more pressing.

Furthermore, hundreds more contracts were set to expire at the end of June 1942 just as work prospects and wages in Newfoundland were greatly improved, giving the overseas men choices previously unavailable to them. The NCG's threat of withholding relief would not keep them where they were, but that was where the UK government wanted them.

Geoffrey Shakespeare's Report

In response to the crisis of morale and discontent within the NOFU that had come to a head by the winter of 1942, Secretary of State for Dominion Affairs Clement Attlee asked Geoffrey Shakespeare, his Parliamentary Under-Secretary, to lead a group of senior officials on an inspection tour of the camps. The participants included Gerald W. St. John Chadwick of the Newfoundland Branch of the Dominions Office, and senior officials of the Home Grown Timber Production Department—Gerald Lenanton, Director; Samuel Strang Steel, Deputy Director; and General Sir Harold Carrington, Assistant Director—along with Jack Turner and Alfred Hart of the NOFU. From February 20 to February 23, the group visited nine camps in three of the four districts where the NOFU operated: Strathmashie, Feshie Bridge, Baillieward, Grantown-on-Spey, Moy, Achanalt, Douniehill, Rosemarkie, and Newtonmore. At each one, they spoke with workers, as well as with foremen and superintendents (Shakespeare, 1942). Following the tour, the group met at Inverness to solidify impressions and build consensus on recommendations. There was unanimity about what needed to be done.

Shakespeare's report to Attlee was an indictment of the conditions that the Ministry of Supply had ignored in relation to its Newfoundland workers.

It included several recommendations for change and was a turning point for both the welfare of the NOFU as well as its reputation while in the UK. While acknowledging that, given the isolated location of its camps, providing amenities for the NOFU was a task of "exceptional difficulty," Shakespeare underscored that officials in the Ministry of Supply, whose Director had been resistant to improvements, must make "a special effort to remove grievances," given what he called the "questions of Imperial policy" that were involved (Shakespeare, 1942, p. 1).

Shakespeare provided several recommendations for the men's welfare and well-being: appointing a headquarters-based welfare officer to oversee unit welfare implementation and arrangements; appointing four welfare officers—one for each district—to administer district-level welfare; building recreation huts and canteens and providing materials for diversions such as books, magazines, and games; and providing free transportation to nearby towns for leisure and religious purposes. Shakespeare noted that the costs of these recommendations needed to be weighed against the overall benefits: "Whatever the cost of making these elementary welfare provisions, it is probable that the improvement of the morale, spirit and happiness of the loggers will result in increased output" (Shakespeare, 1942, p. 4).

Attlee urged Minister of Supply Andrew Duncan to act on the recommendations, noting: "As you know, our relations with Newfoundland are very special at the present time and it is most important not to do anything which might lead to ill feeling against this country" (Attlee, 1942a, pp. 4–5). Now that economic prosperity was returning to Newfoundland, the question of its future political status was pressing on the UK government. The *Amulree Report* had clearly indicated that, once the country was self-supporting, "responsible government, on request from the people of Newfoundland, would be restored" (NRC *Report*, 1933, p. 197). A diplomatic mess involving thousands of Newfoundland men working overseas in support of the war effort would create an outrage best prevented.

Despite the agreement at Inverness and Attlee's positive response to the report, changes did not proceed quickly. During a May 1942 visit to several of the camps,[39] Parliamentary Under-Secretary of State for Dominion Affairs Paul Emrys-Evans found that none of the recommended changes had been made. Canteens had not been built, and petrol had not been made available for recreational visits to nearby towns (Emrys-Evans, 1942a). In a meeting during that visit, the NOFU superintendents emphasized that unless unit grievances were addressed, the men whose contracts were due to expire in the coming weeks would not renew. Furthermore, the superintendents themselves were reconsidering their own commitment to the unit. They shared with Emrys-Evans that they felt the men of the unit were neglected by the NCG and had been "treated without consideration" by the Ministry of Supply (Emrys-Evans, 1942a, p. 1). Emrys-Evans wrote to Attlee that the situation was "clearly intolerable" and "[b]y neglect and lack of understanding on the part of the Ministry of Supply . . . we are well on the way to creation of an absolutely unnecessary difficulty and too much ill feeling in Newfoundland" (1942a, p. 1).

Weeks later, Charles Peat, Parliamentary Secretary to the Minister of Supply, also visited the camps. He was accompanied by those who had visited with Geoffrey Shakespeare—Lenanton, Strang Steel, and Carrington of Timber Production and St. John Chadwick of the Dominions Office. Peat met with Turner, Alfred Hart, and Joseph Curran of the NOFU, as well as superintendents, welfare officers, and NOFU members. His report to Minister Duncan frankly laid out his observations. Peat's impression of the camps and the men he met was "on the whole a good one," but he commented that "there did not seem to be very much initiative among the men or their leaders towards improving conditions which were under their control" (Peat, 1942b, p. 1). Revealing a telling (and inaccurate) assumption, he added: "The atmosphere was the same, I presume, as would exist in the temporary lumber camps erected in Newfoundland" (Peat, 1942b, p. 1). Nor did Peat see anything amiss in the selection of welfare officers, although he noted that the men themselves thought they were not qualified for the job they had been asked to do. Peat's "good enough" attitude clearly failed to take into account the impact on the men of separation from their homes, which had been

39 Emrys-Evans visited all camps except those on Deeside and Rosemarkie on the Black Isle.

compounded by public and government criticism at home and in the host country, leading to poor morale and large-scale resignation from the unit.[40]

Peat did acknowledge, however, both the tense relationship between the NOFU and the Home Grown Timber Production department and its unfair treatment of the unit, the latter somewhat a result of the former. He conceded that "trouble arose owing to the worst possible start when the unit first arrived in England" and after which relationships "steadily deteriorated" (Peat, 1942b, p. 1). As a result of "having got a bad name at the beginning," the unit was thus treated with "annoyance and suspicion" by Timber Production (Peat, 1942b, p. 1). In addition, confidential correspondence between the Ministry of Supply and the Dominions Office concerning the NOFU was sent uncensored to Governor Walwyn and the NCG—a practice Peat suggested should stop immediately. Such correspondence, unintended for the NCG, contributed to the impression in Newfoundland that the unit was both an embarrassment and a failure.

Peat also reported that the men themselves felt they had "suffered a gross injustice," because from their perspective their perceived failure had arisen from conditions over which they had no control (1942b, p. 3). He acknowledged that "there is an element of truth in this" and, as redress, recommended a letter be written to the Dominions Office to reset this impression, one that offered an account of the history of the unit, the initial difficulties it encountered and how they had been surmounted, and how the unit was now involved in "first-class work" (1942b, p. 3). Such a letter was subsequently prepared and sent on June 26 to the Dominions Office by Andrew Duncan, Minister of Supply, and forwarded to the NCG in mid-July by Under-Secretary Eric Machtig. Machtig wrote that the Dominions Office, now in damage control, hoped the report would "serve to remove certain misapprehensions which may have been felt in Newfoundland as to the value which we attached to the work which members have been doing in this country" (Machtig, 1942).

---

40 See Kelly & Forsyth (2018) for accounts of the atmosphere of woods camps in Newfoundland and the camaraderie that was a hallmark of woods camp life, one that provided a buffer against the effects of hard work and isolation.

The concerns raised about the welfare conditions of the NOFU were set in relief by the different treatment UK officials accorded the NOFU and the Canadian Forestry Corps, who operated in nearby camps. Two years into their work in the UK, the NOFU was still arguing for amenities and privileges that the Canadian Forestry Corps took for granted: "a high standard" of welfare provisions, including large recreation huts and attached canteens, and the provision of weekly transportation to nearby towns (Shakespeare, 1942, p. 4). The double standard was not lost on Clement Attlee, who wrote to Minister Duncan: "The position of the Newfoundland Unit should in all essentials be the same as that of the Canadian military forestry unit.... Emrys-Evans tells me that the contrast ... as regards contentment and amenity is most remarkable" (Attlee, 1942b, p. 4).

Canteens

As early as March 1940, just three months after the arrival of the unit in the UK, the Church of Scotland had raised concerns about welfare services for the NOFU. The Church's Canteen Committee at Hawick expressed a sense of responsibility to attend to the NOFU men who visited their community every other Saturday: "Like so many of our boys, they are engaged in national service, and have come from afar to render that service. Hence we are duly bound to do for them what is being done for others similarly situated among us" (Finch, 1940, p. 4). By late 1940, the Church had appointed the Reverend S. J. M. Compton of Tolbooth St. John's Church, Edinburgh, as Deputy to the NOFU ("Welfare," 1940).[41]

By late 1941, the Church of Scotland had opened and was operating a canteen at Camp 6, Laggan (Shakespeare, 1942). Plans to operate additional canteens were slow to unfold as the unit awaited Timber Supply's approval to build recreation huts to accommodate them (A. E. Hart, 1941a). In August 1941, plans for Church of Scotland canteens were thwarted by an order from the War Office that banned civilians from entering them. Protesting the order, Turner underscored to the War Office that the ban "has been felt very acutely throughout the Unit, and, particularly in remote parts of Scotland

41 Compton was a chaplain with the Canadian Mounted Rifles in the First World War.

where small camps are established, it has resulted in severe hardship to the men who have no other recourse to social amenities" (Workman, 1941, p. 1).

In January 1942, Roger Orr, Honorary Secretary of the Scottish Branch of the Empire Societies War Hospitality Committee, expressed his ongoing concern to its London office regarding the welfare of the NOFU. He indicated to Organizing Secretary Dorothy Thompson that the difficulty for the NOFU was their civilian status and that "the responsibility [for the NOFU] is really with our Forestry Commission and I am not at all satisfied that the men are really getting a square deal" (Orr, 1942, p. 1). Orr expressed "a very unpleasant and dissatisfied feeling" about the men's welfare, as well as the need, politically, to leave them with "some kind of decent feelings for this country" (Orr, 1942, p. 2). Orr concluded that an investigation into the conditions of the NOFU by the Dominions Office was necessary so that measures could be taken to address the needs of the men.

Such concerns underscored the ongoing need for welfare services that would make the NOFU more self-reliant. In the latter part of 1941, the NOFU, through its administrative assistant, Alfred Hart, pursued with Timber Supply the issue of establishing and staffing dry canteens in the camps. Hart pointed out that what this would mean to the men was far more than simply offering the pleasures of a canteen:

> The psychological effect on men, returning from a hard day in the woods, covered in mud and possibly wet through, who know that (after a meal) there is a well-lighted and warm recreation hut, in charge of a cheery individual, where they can write letters, read, play games, listen to the radio, and purchase tea, food, cigarettes, etc., etc., should not be overlooked. Many men would remain in camp and spend a happy evening with their pals instead of meandering about as at present and, if in funds, being compelled to seek solace in the nearest public house. (A. E. Hart, 1941c, p. 2)

Hart's efforts on behalf of the unit, as well as those of Turner, had little impact on the pace at which Timber Supply addressed the issue. The isolation and lack of amenities struck Geoffrey Shakespeare and his party during their visit and the need for canteens was a central recommendation in his report, yet Paul Emrys-Evans found no work had been done in response, noting that "without canteens or any other social life it is remarkable that the morale of the men has not deteriorated" (Emrys-Evans, 1942a, p. 1). By the time of Peat's visit in June 1942, and no doubt as a result of the appeal by Emrys-Evans to Attlee (and his subsequent action), construction on many of the canteens had begun or was nearing completion. A solution for how to staff them was close at hand: the local wives of NOFU members were hired to manage the canteens, and living quarters were provided for these couples—one per camp—in an annex of the recreational building (NOFA, 1945).

The canteens were an essential addition to camp infrastructure. They reduced the austerity of camp life by providing some of the amenities of the "vans" that were a key feature of camp life at home, but with the added bonus of after-hours light meals—"tea and a bun or sandwich"—all while remaining within the ration structure approved for the unit on March 1, 1942 (A. E. Hart, 1942a, p. 1). Canteens also offered billiards and other games, a radio, and reading materials, including newspapers from home (NOFA, 1945), and were a venue for concerts and films provided by the Entertainments National Service Association. St. John Chadwick reported that the canteens provided "splendid trade in everything from braces to chocolate bars and ... that extra something which makes a change from the meals served in the dining halls, good and copious though these are" (St. John Chadwick, 1944, p. 5).

Sir William Brass, Conservative Member of Parliament for Clitheroe, visited NOFU camps several times from 1942 onward. Brass chaired the Newfoundland Hospitality Committee, a parliamentary committee concerned with the welfare of Newfoundlanders in the UK. He also served as Chair of the British Film Institute during the war and photographed the NOFU men in moving and still pictures, some of which accompanied articles publicizing the work of the unit. Brass led Friday morning tours of Parliament for members of the unit visiting London. In a letter to his constituency's local newspaper following a springtime visit to NOFU camps in 1944, he described the men as "real experts" who were doing "excellent work" and were "anxious to see the job through to the end" ("With the Loggers," 1944, p. 2).

Petrol

The rationing of petrol meant there was little available for anything but wartime essentials. At NOFU camps, petrol was reserved for work purposes—to operate lorries and tractors, for example, or to make work-related trips. A request for an additional allotment so that men could access non-work-related weekly visits to nearby villages and towns was contentious. Petrol use was closely monitored and the unit sometimes found itself under scrutiny, with locals complaining to authorities about perceived misuse.[42]

Following his visit to the NOFU camps, Geoffrey Shakespeare recommended an allotment of 100 gallons per week to the unit "for divine service and for recreational purposes" (Peat, 1942c, p. 1). As Shakespeare noted, many of the NOFU camps were in remote districts that lacked any timely or convenient local trains—and private car hire was costly. Without access to villages and towns, he noted, "the camps become internment compounds" (Shakespeare, 1942, p. 3). Furthermore, the Canadian Forestry Corps had already received such concessions of petrol for its recreational use.

Shakespeare's recommendation resulted in a flurry of correspondence between various departments of the Ministry of Supply, who took a consistently hard line on any requests from the NOFU. It often offered the rebuttal that an exception to one group would discriminate against all others and place the Ministry in an impossible position. It seemed lost on the Ministry of Supply that the NOFU was already treated exceptionally—but in the worse possible sense. Despite what the Dominions Office believed were "strong reasons of Imperial policy" (Fitzgerald, 1942b, p. 1) to make this exception, the Ministry of Supply, after consultation with the Petroleum Department, conceded on petrol for access to divine service only, adding that "we will not press them on this point" of recreational access due to orders that "all pleasure motoring must cease" (Sears, 1942, p. 1).

42 See M. J. Taylor (1942) for a discussion of complaints about petrol use at Rosebery camp.

Attlee refused to take no for an answer and involved more senior officials on this issue. In late August 1942, he wrote Gwilym Lloyd George, Parliamentary Secretary at the Ministry of Fuel and Power, urging him to concede on the topic of petrol. Attlee wrote:

> I feel that to refuse this privilege would have results more disastrous to production than would at first sight seem apparent. . . . [I]t is possible to justify special treatment for this Unit, first because it is an overseas body and we owe it to the Newfoundland Government, who gave such prompt assistance in the early stages of the war in the recruitment of these men, to keep the loggers as contented and well provided for as possible; secondly because the Canadian, Australian and New Zealand Forestry Units working here under similar conditions have always had a sufficient supply of petrol to permit their men to visit their nearest centres when off duty. (Attlee, 1942d, pp. 2–4)

Lloyd George agreed, stipulating only that strict controls be developed for its use (Lloyd George, 1942). By mid-October, Alfred Hart had distributed a lengthy list of conditions for lorry and petrol use to the unit (A. E. Hart, 1942c). It included various stipulations regarding the number of trips, maximum distance from camp, purpose and timing of trips, maximum number of men travelling, lorry drivers, and lorry parking. Misbehaviour was punishable by revocation of travel privileges. Welfare officers would supervise arrangements for travel and oversee compliance with the rules, with the final authority for approval resting with superintendents and the Officer-in-Charge (A. E. Hart, 1942c).

Wages

Considering changes in economic conditions as the war progressed, including increases in the cost of living in both countries, Officer-in-Charge Turner sought a wage increase for the unit in the late fall of 1942. The NOFU men had left home with a contract that reflected wages in Newfoundland at the time—but wartime prosperity and labour shortages at home had driven wages upward in their industry. Wages had also increased in the UK, yet the NOFU wages remained static As well, wartime conditions had increased the cost of staples, so remittances sent home were falling short in relation to the cost of living in Newfoundland, which would increase by almost 60 percent during the war (High, 2003).

Consequently, wage discrepancies became a factor in non-renewal of contracts. Some workers resigned to seek better opportunities at home, where the pulp and paper companies were under pressure to maintain production despite their own dire labour shortage, estimated to total three thousand loggers. In a country-wide radio broadcast on June 28, 1942, Commissioner of Natural Resources Dunn issued an appeal for loggers ("Loggers Wanted," 1942). Hoping to alleviate similar labour woes, Canada raised its shortage of loggers in a discussion between the two countries ("Canada Is Seeking Loggers," 1943). The situation there was so dire that it resulted in the repatriation of several companies of the Canadian Forestry Corps.

In their May 1942 meeting with Emrys-Evans, the NOFU superintendents had flagged the issue of pay parity (NOFU Superintendents, 1942). Turner initially requested a 25 percent across-the-board increase for the unit. This amount was rejected by the Ministry of Supply on the basis that existing wages and subsistence combined already placed NOFU workers in a better position than UK labourers. Instead, effective November 1942, the Ministry offered the following wage increases: Loggers on piece work, $2.50 per week; loggers not on piece work and other workers, an increase of six cents per hour; foremen, an extra $5 per week; and superintendents an additional $7 per week. Turner accepted, and the NCG approved the offer (Attlee, 1942f; "Increase in Pay," 1943).

Although the Dominions Office had granted that the "increased cost of living in Newfoundland justifies Ministry of Supply increasing existing rates to permit men sending home higher remittances" (Tait, in Attlee, 1942f, p. 1), Turner was able to negotiate some leeway on members having to send more money home. Turner agreed to "influence his men" to increase their allotments and, in return, the Ministry of Supply would not require as a condition of payment any proof of an increased allotment (Tait, in Attlee, 1942f, p. 1).

These increases gave the NOFU men a little more earning power, and with the quiet option to spend it in the UK, to invest it in War Savings Certificates or send it home to family. It was the only wage increase the unit would receive during its UK employment.

Administrative Tensions: Reasserting Leadership and Autonomy

In the first years of the NOFU operations, two levels of strain arose from its treatment by UK officials: within the rank and file, this stress was related to their living conditions and overall treatment; for the Officer-in-Charge, it related to vested authority and chain of command. The Dominions Office tried to mediate those tensions while also managing its own relationship with the NCG, all while avoiding any further strain that might have potentially difficult consequences.

The tensions related to the NOFU existed in a context of managing imperial relations, which were themselves related to the removal of self-government in Newfoundland, the unpopularity there of the NCG, and the growing question of how future governance would be resolved as economic conditions improved in the colony. The Dominions Office therefore had an increasingly strong interest in the unit, "since any dissatisfaction on the part of its members is at once reflected in Newfoundland" (Emrys-Evans, 1942c, p. 2). As it wrestled with the Ministry of Supply about amenities for the men, the Dominions Office also had to deal with the consequences of its attitude toward the Officer-in-Charge and the legislative authority by which he was appointed.

Under the legislation by which the NOFU was formed, control and management of the NOFU rested with Newfoundland's Commissioner of Natural Resources and the unit's Officer-in-Charge (*An Act*, 1939). In the UK, first Timber Supply and, from 1941 onward, Timber Production of the Ministry of Supply dictated all matters related to production—timber stands, camp locations, production specifics, etc.; all other matters fell under the purview of Turner and his NCG superiors.[43] The chaotic start for the NOFU under Timber Supply and its unrelenting criticisms of the unit set the tone for difficulties that shaped the NOFU experiences thereafter. These difficulties included criticism of Turner's leadership and a general undermining of his authority. According to St. John Chadwick, the legislative terms of An Act Respecting the Newfoundland Overseas Forestry Unit was "held tacitly in suspense by the Ministry of Supply," and Turner, perhaps wanting to avoid conflict in order to focus on timber production, had not "seen fit to stand on the letter of the law" (St. John Chadwick, 1942c, p. 9).

Gerald Lenanton, Director of Timber Production, was a candid critic of Turner, at one point complaining about the Officer-in-Charge's practice of approaching both Timber Production and the Dominions Office on various issues. This "dual approach on so many problems," Lenanton argued, "does not ease, and certainly does not quicken procedure, or make for good working" (Lenanton, 1942a, p. 1). In a meeting with the Dominions Office earlier in 1942, Lenanton was much more frank, stating that Turner was "not a big enough man for the job and lacked qualities of leadership essential to operations on a considerable scale" (St. John Chadwick, 1942b, p. 1). The tension was obvious to officials in both countries and was documented in the competing versions of the unit presented in cover letters to Turner's reports and Turner's responses to them. On an earlier visit to the NOFU camps (spring 1941), Commissioner Penson suspected there was "some prejudice" against Turner within Timber Production (Penson, 1941a, p. 2). Penson offered a weak defence of the Officer-in-Charge, stating "he is not a super-man," and, adding his own undercutting of Turner, "perhaps not even the best we could have chosen . . . had we known beforehand the particular kind of difficulties involved" (Penson, 1941a, p. 2).

Given that he faced disregard for the legislation that constituted the NOFU, Timber Production's consistent unresponsiveness to the concerns of his men, and both the NCG's and Timber Production's disregard for the unit's leadership, Turner was left little choice but to proceed with simultaneous requests on

43 The NOFU was the only overseas forestry unit in Scotland to work with the Timber Supply division of the Forestry Commission; the Home Grown Timber Production Department was the governing UK authority by the time other units arrived.

two fronts. The Dominions Office offered a more favourable hearing than Timber Production, which Paul Emrys-Evans accused of "ineptitude, incompetence and lack of sympathy" toward the NOFU; and all that, he felt, was worsened by the NCG's neglect of the unit (Emrys-Evans, 1942a, p. 7). Undoubtedly, Turner's task was huge—to oversee and manage a large number of civilian men scattered across a foreign country while meeting the expectations of UK authorities regarding production (to be achieved in difficult terrain and weather). But he also lacked reasonable support for his unit, whose status was undermined and treated as subpar by Timber Production and for whom every issue was met by an elaborate bureaucratic wall of delay, detraction, or denial.

Circumstances were much different for the Canadian Forestry Corps, for whom "no serious problems resulted" (Wonders, 2008, p. 22) from its own dual oversight situation—Timber Production and Canadian Military Headquarters, London. That unit was self-contained, well-equipped, and fully supported by the Canadian government. As Emrys-Evans noted, "it was clear that no interference from Timber Control would be tolerated [by the Canadian Forestry Corps]" (Emrys-Evans, 1942b, p. 2) or, it is assumed, even attempted. The difference in their experience and treatment was related not only to the personnel involved and the status of the units—civilian versus military—but to the status of the dominions themselves at the time. In no small way, it also reflected the lack of action by the NCG on behalf of the NOFU and its leadership.

The relationship between the NOFU and Timber Production was so unsavoury that it threatened the entire leadership structure of the NOFU. When Emrys-Evans arrived at Newtonmore in May 1942 for what could only be termed an emergency visit, Turner advised him that, owing to Timber Production's interference in his work and the humiliation he had suffered at its hands, he would find it hard to continue in his position (Emrys-Evans, 1942c). The trouble arose, in part, from what Emrys-Evans described as "an astonishing attitude of mind on the part of [Director Lenanton]" that was so prejudicial that, in the opinion of Emrys-Evans, Lenanton was "plainly quite unfitted" to report the doings of the NOFU to the NCG (Emrys-Evans, 1942c, p. 6).

Lenanton's lack of respect for the unit and its leadership had repercussions down the chain of authority in Timber Production. Assistant Director

General Sir Harold Carrington revealed to Emrys-Evans that he had no knowledge of the Act that guided the formation and authority structure of the NOFU. Carrington pointed out that, had he known the correct status of the unit and its Officer-in-Charge, he would not have given instructions to Turner in the manner in which he had, as if Turner was a subordinate, similar to a divisional officer in the Home Timber units (Emrys-Evans, 1942b).

By 1942, it was clear that the Ministry of Supply oftentimes did not consult with Turner or apprise him of policy and its implementation that were related to the NOFU. Timber Production sent directives to NOFU superintendents and foremen, bypassing Turner (Emrys-Evans, 1942e). Turner himself was told that his correspondence with the Dominions Office and the NCG should first be passed to Timber Production (Emrys-Evans, 1942b). Timber Production reached into the management and authority structure of the unit in ways that far exceeded its own authority, which rested solely in matters related to timber production.

Its unilateral decisions included moving the NOFU headquarters to the Timber Production offices for "closer liaison" in February 1942 (Lenanton, 1942b, p. 3). At that time, Timber Production's General Accounts branch also subsumed the NOFU accounts division to reduce duplication following a complaint by Lenanton to the Dominions Office earlier in the year (St. John Chadwick, 1942b). Catering and welfare for the unit were also placed under the control of General Carrington (Peat, 1942a). It was an administrative coup that wrested control from the NOFU leadership and moved it to the Ministry of Supply.

The Superintendents' Document

Timber Production interference reached a point where, in a special meeting with NOFU superintendents and Emrys-Evans during his May 1942 visit, they presented him with a two-page document outlining their perspective on the state of the unit. In it, they made numerous claims related to its status, the undermining of Turner's authority and credibility, and the well-being of the men. They pointed to questionable relations between Timber Production and some estate owners, which had resulted in decisions and practices that were inconsistent with official statements about timber shortages. This nepotism, they claimed, had affected timber and equipment allocations, and compromised UK national interests as well as the best interests of the unit, which had sometimes been forced into premature relocation.

The superintendents claimed that the setup of the NOFU in the UK bore little resemblance to the original plan, as set out in NCG legislation governing the unit (*An Act*, 1939). They reiterated their "absolute confidence" in the Officer-in-Charge and his ability "to secure maximum efficiency of the unit" (NOFU Superintendents, 1942, p. 1). The superintendents also addressed the cold and uncooperative attitude of Timber Production toward the NOFU. They were candid about the implications of changing economic conditions—including rising cost of living—for its members and their families; and specifically criticized the wage rates the men were paid while in the UK, where they could access neither the comforts of home nor the privileges offered the armed forces. They concluded that none of the authorities under which the unit operated—the Ministry of Supply, the Dominions Office, and the NCG—appreciated or understood the challenges the unit faced (NOFU Superintendents, 1942).

The superintendents felt that the NCG had abandoned its interest in the unit and the men. They noted that the little means by which the NCG learned about the unit's work—the reports from Turner—were "construed by [UK government officials] to present their opinions" (NOFU Superintendents, 1942, p. 1), thereby distorting the conditions, challenges, and achievements of the unit. Particularly offensive, in this regard, was a cover letter by the Director of Timber Production to Turner's report covering the period from September 1941 to January 1942, in which Lenanton attempted to diminish the impressive production rates of the NOFU in comparison to Scottish home labour. Rather than commending the NOFU for its work, Lenanton downplayed the accomplishments of the NOFU, using the tropes of physical strength and experience so well-publicized about the men to suggest that these features somehow gave the NOFU an unfair advantage and thus nullified Turner's comparison (Lenanton, 1942b). This attitude infuriated Emrys-Evans, who wrote that "it is as if when Dominion troops had done well, the War Office had attempted to explain their success away" (Emrys-Evans, 1942b, p. 6).

To remedy the situation they described, the superintendents offered several requests and suggestions. First and foremost was compliance by UK officials to the Act governing the unit, including recognition of the authority of the Officer-in-Charge; NOFU leadership should be subject only to the jurisdiction of the Dominions Office and the NCG. They suggested the Governor in Commission appoint all unit officers, as recommended by Turner. They requested a reconsideration of wages in the context of changing conditions at home, and a review of the unit by the Dominions Office, including contracts and compensation, issues of discipline, and opportunities for the involvement of NOFU administrative staff in matters related to its workings. Finally, the superintendents suggested that if these things were not possible then the Dominions Office should consider transforming the unit into a military or semi-military unit similar to the Canadian Forestry Corps, and that it should be led by Turner, an ex-army Captain (NOFU Superintendents, 1942).

The document was a stunning rebuke of both Timber Production and the Ministry of Supply, their prejudicial attitudes and actions in relation to the NOFU, and the indifference of the NCG toward the unit. It was also a gesture of faith in the Dominions Office and their ability to address the problems outlined. Emrys-Evans was impressed by the superintendents and how they presented their case "with a moderation and sense of public duty which in every way does them credit" (Emrys-Evans, 1942c, p. 7). He was also aghast at the behaviour of Timber Production, in particular that of its Director, Gerald Lenanton. Following on the superintendents' requests, Emrys-Evans called for immediate action to reset the role and authority structure of the NOFU, to inquire into the failure of Timber Production to respect the Act governing the unit and follow through on agreed actions to improve its conditions, to investigate increased pay for NOFU members, to clarify lines of communications between the NOFU and the Dominions Office and NCG, and to explain to the NCG what had occurred, including a request that they improve publicity for the unit at home (Emrys-Evans, 1942c).

In a letter to Clement Attlee following his visit, Emrys-Evans railed against the Ministry of Supply over its attitude toward the NOFU. He reported that he had informed General Carrington that they "had exceeded their powers and that I was determined some head should fall" (Emrys-Evans, 1942a, p. 1). Clearly perturbed by what he had learned about Timber Production, Clement Attlee urgently outlined the problems to Minister of Supply Andrew Duncan. Attlee wrote to Duncan that Timber Production was "ignoring the status of the Unit as an overseas Unit raised by the Newfoundland Government under its own officer-in-charge . . . and have been treating it precisely as a Divisional section of their own organization" (Attlee, 1942b, p. 2). Attlee's comment highlighted the importance of the word *overseas* in the name of the unit, especially in the UK. Attlee cited the egregious evidence:

> This [attitude] has resulted in instructions being issued to camp foremen and superintendents over the head of Captain Turner, and I have seen communications to Captain Turner from the Director of Home Grown Timber [Gerald Lenanton] stating in effect that [Turner] has no voice in questions of the staffing and organization of the Unit. This is entirely contrary to the provisions of the Newfoundland Act which delegated the control and administration of the Unit through the Governor in Commission to Captain Turner. (Attlee, 1942b, pp. 2–3)

In one of these offensive staffing actions, Timber Production had run roughshod over the authority of the Officer-in-Charge when appointing a welfare officer for the NOFU. The position had been approved in fall of 1940 and a recommendation for hire was made by Turner, but Timber Production failed to act on it. When the appointment was finalized in 1942, the decision was made unilaterally by Timber Production, in what St. John Chadwick called "a very bad move" made hastily under pressure after a follow-up camp visit by officials. Furthermore, the appointee, Major W. K. Henney, was unsuitable, someone St. John Chadwick considered "far too much a man about town to get on well with Newfoundlanders, and one could not imagine him starting a song or joining in camp life with any gusto" (St. John Chadwick, 1942c, pp. 28–29).

Charles Peat reported that the NOFU men were disgruntled with the appointment of Henney and had shared with him "a strong feeling" that the appointment should have been discussed with Turner (Peat, 1942b, p. 2). Turner had recommended H. G. Penfold, formerly a senior education and welfare office with the Ministry of Labour, a man Turner commended for having "an excellent reputation" and, as a pianist and piano-accordionist, having "a remarkable knack of getting men together for a sing-song and making them feel at home" (A. E. Hart, 1940, p. 2).[44]

Attlee was clear about the action he expected to redress the "unsatisfactory business" between Timber Production and the NOFU, noting that while the role of the former in matters of timber production and allocation of the unit was indisputable, there should be no doubt about who was in charge: "[T]he internal organization, administration and discipline of the Unit must under the Newfoundland legislation rest with its own officer-in-charge and psychologically also it is essential that this should be so" (Attlee, 1942b, p. 3).

The matters were not only ones of timber production and clear lines of responsibility. Attlee reinforced to Duncan the broader importance of resetting relations with the unit and its leadership:

> You will, I know, appreciate that our relations with Newfoundland are involved in the treatment of these men, who have come overseas at considerable personal sacrifice and with patriotic motives to help in our own war effort. They are doing splendid work and have recently broken all records for timber cutting operations for any unit in Scotland. I think that they have not in fact been given sufficient credit for this and they made it clear to Emrys-Evans that they feel this. All the more reason therefore why the question of the status of the Unit should be put on a proper basis. (Attlee, 1942b, pp. 3–4)

---

44 When the welfare section was placed under the administration of the NOFU in late 1942, Penfold was eventually assigned to such a position and remained in it until the end of the unit's time in the UK.

The men of the NOFU, as well as its leadership, were unaware of the extent to which their well-being was now part of a diplomatic context related to the unfolding of the larger question of the status of their home country.

The tensions between the Ministry of Supply and the Dominions Office around the treatment of the NOFU resulted in a note of agreement, signed on June 29, 1942. The "Heads Agreement," as it was called, addressed four key sources of friction: the NOFU organization, including its Accounts branch, was reverted to its previous separate administrative offices; all correspondence from Timber Production related to the NOFU—instructions, memoranda, etc.—was to be sent directly to Turner; the equipment branch of the NOFU was merged with that of the Timber Production branch; and the Ministry of Supply remained responsible for welfare organization, including canteens, recreation huts, and welfare officers, despite its request that these responsibilities pass to the Dominions Office. The agreement underscored that Turner "must be kept fully informed of all matters of policy and procedure" related to the NOFU (Peat, 1942c, p. 2). The Dominions Office accepted the note of agreement, with Emrys-Evans reiterating that Turner was "in sole and complete charge of the N.F.U. and all matters connected with it, subject only to the overriding authority of the Home Grown Timber Production Department in directives relating to timber cutting operations" (Emrys-Evans, 1942c, p. 1).

Following the signing of the agreement, matters related to authority and welfare moved much more quickly. Canteens were completed, educational plans for the unit began to roll out, and the lines of communication between the NOFU, the NCG, Timber Production, and the Dominions Office were clarified. The Timber Production Director's practice of adding cover letters to Turner's reports to the NCG was discontinued. Turner was reassured that it was within his authority to communicate directly with the NCG and the Dominions Office. As for Lenanton, Emrys-Evans wanted questions answered about why the NOFU had been treated as it had and warned that Lenanton, in the meantime, "should not enter into any further tactless controversies" (Emrys-Evans, 1942b, p. 8).

As tensions abated, another round of visits to the NOFU camps began. In late August 1942, Emrys-Evans, Admiral Bromley, and St. John Chadwick of the Dominions Office conducted a follow-up visit to all camps visited in the spring. Emrys-Evans reported "an entirely different spirit" from that which he had observed during his May visit (1942d, p. 1). Turner and the superintendents shared that they felt their grievances had been heard and addressed and Timber Production had, after Peat's visit, treated the NOFU with greater consideration, in recognition of its status as an overseas unit. Turner also reported that far fewer men (than the three to four hundred anticipated before the welfare changes were made) had returned to Newfoundland upon expiration of their contracts on June 30, and the eighty or so who had left had done so for what he considered to be good reasons (Emrys-Evans, 1942d).

Early in September 1942, Secretary and Deputy Prime Minister Clement Attlee visited NOFU camps in Aberdeenshire. In a radio broadcast to Newfoundland made during a visit to the country shortly thereafter, he reported that the NOFU was "a splendid lot of men, fit and healthy." He reported that they were "happy" and "well fed," and that they were "doing a service no less vital to the war effort than those of the fighting men" (Attlee, 1942e, p. 3).[45]

NOFU members lunching with the UK's Deputy Prime Minister Clement Attlee (seated next to the window), at a camp near Ballater, Aberdeenshire, 1942. (Courtesy of the Ballater Historic Forestry Project Association)

45 For Attlee's observations on Newfoundland, the NCG, and the tricky question of Newfoundland's future, see Neary (1984).

Later in the fall of 1942, NCG Commissioner of Defence Lewis Emerson visited NOFU camps in the Carrbridge district as part of a tour that included all overseas services from Newfoundland. Emerson made a broadcast about his visit on the evening of December 8, 1942. In a home-front effort to create more credibility for the NOFU, he acknowledged the disorganization and resulting dissatisfaction and demoralization the unit had experienced in its initial years of work—due largely, he said, to the lack of preparation for their work by UK officials. Emerson skirted the more disturbing details, including inaction by the NCG, explaining only that "a peace-time [UK] government had not yet turned to full war effort." He added that, despite the difficulties, the NOFU was now in "the full peak of production" (Emerson, 1942, p. 4).

Despite these changes, by the end of 1942 the unit lost some of its senior officials, who transferred to other forces or other employment. Joseph Curran, the unit's second-in-command, had returned to Newfoundland late in the spring of 1941 to help recruit the third and final group of foresters. Negotiating his return to Scotland in winter 1942, Curran indicated that he was willing to commit only for a period of one more year, not the duration (Dunn, in Walwyn, 1942b). Curran did return to the UK in the spring of 1942, almost a year after he'd left for Newfoundland. He was repatriated in late June 1943.

While the source of many of the difficulties experienced by the NOFU rested in its treatment by UK officials, the unit's problems were intensified by the indifference of the NCG, which did not intervene on its behalf to protest with any force the conditions under which the men worked. Added to this indifference was the unit's recognition that the NCG also distinguished between those serving overseas in military roles and those serving as civilians, such as the NOFU. Its preferential attitude for the former was clearly evident in how it oversaw the supply of comforts to those who were overseas during the war.

The practice of providing supports and comforts to men and women in service had been remarkably demonstrated by Newfoundlanders and Labradorians in the First World War, and it continued during the Second. The

Patriotic Association of the Women of Newfoundland (informally called the WPA), the Newfoundland Patriotic Association, and the Red Cross were all active in providing a variety of supports—meals, visits, entertainment, accommodations—as well as comforts such as knitwear, clothing, and personal-care amenities to military personnel serving overseas and at home.

Support services and comforts provided to the armed services, however, were not typically offered to civilian volunteers, including the NOFU, although branches of the WPA sometimes sent supplies. In May 1942, R. Travers (NOFU #3164) of Lark Harbour wrote to thank the WPA of his hometown for the packages he and other members of the unit had received. Travers noted that "when we do not receive letters from home we feel rather down-hearted," but packages from the WPA were a reminder that, even if others had forgotten them, they were not forgotten by the WPA (Travers, 1942, p. 9). This policy slight bothered the men of the unit.[46] Sometimes Officer-in-Charge Turner sent a direct request for comforts, such as before Christmas 1942, when he asked for and received from the WPA one pair of woollen socks for each man in the unit.

By Christmas 1943, perhaps as a trickle-down effect of the developments related to the unit in 1942, things had improved. The WPA sent two pairs of socks per man in a total of 1,512 parcels for the unit (Patriotic Association, 1943). But there was a policy catch: the NOFU was not permitted such items without forfeiting UK clothing coupons! The restriction was ultimately repealed and the NOFU men received their packages, but not until Christmas had passed. Packages for members of the 166th Field Regiment (Royal Artillery) that year, meantime, had included woollen comforts, cigarettes, socks, razors, khaki handkerchiefs and playing cards; bathing trunks were even provided for the 59th Heavy Regiment ("Comforts from W.P.A.," 1944).

Some restrictions related to comforts imposed at the beginning of the war were eased as the war progressed. Shortly after Newfoundland recruits joined UK armed forces overseas, in the early months of the war, tobacco and cigarettes were sent excise-free directly from the warehouse at St. John's and, subject to certain conditions, accepted into the UK duty free. In May 1940,

---

46 For an indication of the extent of these services and comforts, see Patriotic Association of the Women of Newfoundland (1943).

a press communique announced that this privilege was extended to cigarettes for the NOFU, with conditions: packages sent were required to have a minimum of 200 cigarettes or a half-pound of tobacco and be addressed directly to a NOFU member in care of the NOFU administrative office in Edinburgh ("Tobacco and Cigarettes," 1940).

On March 20, 1943, Governor Walwyn announced an amendment to the Defence (Control of Export) Regulations. The Regulations prohibited the export of foodstuffs except under strict conditions: small parcels of food could be sent to members of the armed forces serving overseas, in Canada, or in the United States. The amendment added the Newfoundland Forestry Unit to the list of those included in the exception ("Published by Authority," 1943), again a belated change that likely occurred in response to the crisis of welfare in 1942.

While the easing of restrictions was some comfort to the men of the NOFU, it did not seriously challenge attitudes about hierarchies of service that operated throughout and after the war. It would take a more sustained postwar challenge to these hierarchies, in which the ex-NOFU men played no small role, to change the meaning of service and its rewards.

Essential Workers: Another Threat to Autonomy

Amidst the wranglings about the NOFU's welfare, leadership, and status in 1942, the tension around labour retention continued: how best could the unit maintain its numbers? Propaganda was one tool used to encourage men to remain with the unit. Improved camp conditions were another. As the war unfolded, legislation also emerged that affected retention. Under the UK's Ministry of Labour Occupational Classifications, established in 1938, foresters (including loggers, tree fellers, etc.) were listed as members of a reserved occupation. Within this group, only men under the age of 25 were free to enlist in the armed forces (Ewbank, 1940a). Supplementary legislation, the UK Essential Work (General Provisions) Order, was introduced on March 5, 1941. This far-reaching order expanded the list of occupations deemed essential war work and authorized employers to prevent workers from leaving positions covered by the legislation. No such workers could leave or be dismissed from their work without the approval of the National Service Officer.

As early as the fall of 1940, Turner had raised the issue of denying men permission to leave the NOFU for the armed services, citing the possibility of such men then refusing to work (Turner, 1940i). In the summer of 1942, he sought permission to invoke the Essential Work Order within the NOFU "should the necessity arise," to stem the problem of absenteeism from work without cause (Attlee, 1942c, p. 1). The NCG supported this use of the UK legislation (Walwyn, 1942c). That Turner needed it as part of his command arsenal reflects the extent of ongoing disciplinary problems within the unit prior to increased supervision and the resolution of demands for improved conditions in camps.

Secretary of State for Dominion Affairs Viscount Cranborne first raised with Governor Walwyn the application of the Essential Work Order to the NOFU in late December 1941. He conveyed that the Minister of Supply was concerned about the number of NOFU workers who, upon expiration of their contracts (so-called time-expired men), were taking other work in the UK. He also noted that men were leaving the unit in breach of their NOFU contract to take other UK-based work. Cranborne emphasized that the Ministry of Supply could legally apply the Essential Work Order to all members of the unit, but that it would not be applied to those whose contract had expired and who were awaiting repatriation (Viscount Cranborne, 1941c).

In a memo circulated to the NCG following Cranborne's inquiry, Commissioner Dunn asked the other Commissioners to consider this application of the Essential Work Order, pointing out that he saw no objection to the order being applied to workers leaving the unit in breach of contract. Of the time-expired men seeking other work in the UK, Dunn felt that applying the Work Order to ensure that they remained with the unit would be viewed as "conscription of Newfoundland labour applied only to a small number of men who have volunteered, in the first instance, to work overseas" (Dunn, 1941b, p. 1). The other Commissioners agreed and suggested, instead, that a warning be issued to such men: they could potentially be ordered to return to the unit, should they decide to leave (Walwyn, 1942a).

By 1943, the Ministry of Supply was considering a wider application of the order. The woods labour situation in the UK was dire at this point in the war. The British Honduras Forestry Unit was repatriated in that year; the Australian Forestry Group would be repatriated in the fall of the year, and the New Zealand Forestry Companies were to be redeployed to parts of Europe or Africa or repatriated in the fall, as well. In addition, the Canadian Forestry Corps sent several companies back to Canada early in 1943 and redeployed others to the Continent. At this reduced strength and as the NOFU continued to bleed men—in 1942 alone losing over four hundred workers to repatriation, transfers, and dismissals—these two units and the Women's Timber Corps were all that remained to supply UK wartime timber needs. The Ministry of Supply foresaw a further drain on labour over time, so once again looked to apply the Essential Work Order more broadly.

The Ministry of Supply and the Ministry of Labour met and corresponded about this issue throughout the summer of 1943. Despite the agreement reached in 1942 reasserting Turner's authority around all matters related to the unit and requiring all correspondence related to the unit to pass through him (Emrys-Evans, 1942c), Turner was excluded from these discussions. When he finally became privy to the trail of correspondence between W. K. Ward of the Ministry of Labour and General Carrington of Timber Production about a proposed Order in Council for new defence regulations, he addressed the issue of his omission in a lengthy letter to Paul Emrys-Evans. In it, Turner reasserted his authority and that of the NCG around the status of the NOFU contracts, which Timber Production wished to override through the Essential Work legislation. Turner suggested both ministries had overstepped their bounds, once again making decisions about the unit without consultation (Turner, 1943b).

The slight to Turner aside, the UK government had no actual obstacles to applying the Essential Work Order to any man who considered working in the country in other capacities outside the NOFU—all clearly fell under the legislation. As one NOFU worker put it, "they froze us in our jobs in the woods," leaving men without access to other, better paid employment or other, better-supported forms of service ("Foresters Petitioning," 1966, p. 7).

Request for Uniforms: A Quest for Legitimacy

There was also frustration with efforts to ease some of the disgruntlement resulting from the unit's position. Their civilian, un-uniformed status was an issue for the NOFU men throughout their time in the UK. It factored into how the men saw themselves as well as how others saw them and evaluated the status of their work service. Like the suits the men acquired for evenings out in the towns and villages near their camps, uniforms were also tied to a version of wartime masculinity—a man in uniform was a "real man," was the attitude—for which it was a necessity. Without it, as was suggested by the rationale for uniforms, the men were subjected to disparaging remarks related to manhood, patriotism, and service (Emrys-Evans, 1942e; A. E. Hart, 1942b).[47]

Turner's request for a walking-out uniform—supported by 90 percent of the NOFU members—was under review during the height of the welfare and leadership crises in 1942. He proposed a battledress-type of suit in brown, chestnut, or other earth tone, with matching beret, shoulder titles bearing "N.F.U.," and sleeve flashes with "Newfoundland" in red letters (A. E. Hart, 1942b). The proposed style resembled that of the British Honduras Forestry Unit, whose uniform had been issued upon arrival in the UK following the loss of their personal clothing items during a U-boat attack enroute, as well as official concern for men from a tropical area acclimatizing to a cold environment (Lenanton, 1942a).

Alfred E. Hart, NOFU Administrative Assistant, outlined the benefits of a uniform in a memorandum to Assistant Deputy of Timber Production General Sir Harold Carrington in late September 1942. Hart stressed that the proposed uniform would not be worn for work, and it would bring "greater contentment and a happier atmosphere" throughout the unit, which

---

47 For a discussion of masculinities and service in the UK during the Second World War, see Robb & Pattinson (2018).

in turn would result in greater productivity (A. E. Hart, 1942b, p. 1). A uniform would also allow members privileges while travelling to which they were not, as civilians, allowed access.

At the same time, Turner made the case for a NOFU walking-out uniform to the Dominions Office, noting that his un-uniformed men were embarrassed by how troops stationed in Scotland treated them. Under-Secretary Paul Emrys-Evans summarized Turner's position in a letter to Charles Peat at the Ministry of Supply. Emrys-Evans supported Turner's request, adding:

> From personal discussion with the men I know that this feeling [of embarrassment] is by no means exaggerated and that there have been encounters between the loggers and the troops on various occasions. If some kind of uniform was provided, the men would be better treated in public places in Scotland and their self-respect would increase accordingly. It would also create more esprit de corps among the men and would make a good impression on their people at home when they were repatriated at the end of the war. (Emrys-Evans, 1942e, p. 2)

The harm cut deeper than embarrassment. Ex-NOFU men writing to the editor of *The Daily News* in St. John's[48] addressed local criticism of those who had returned, putting it this way:

> The men who came back were not wounded by bullets or shellfire. But there are other ways to be wounded that can be explained if necessary, and the most of us are ready again to stand by the old flag for the cause of the Empire, but it won't be in the forestry [unit]. (Boys of Newfoundland Overseas Forestry, 1940, p. 4)

The Duke of Devonshire, during his visit of April 1940, had also noted that "wounding remarks had been made about the non-combatant nature of the [NOFU] service" (Duke of Devonshire, 1940, p. 1). They included comments

48 John S. Currie was editor of *The Daily News* from 1916 to 1956.

that the men found especially offensive, made during a by-election in Argyllshire by William Power, a candidate for the Scottish Nationalist Party.

There were also reports of NOFU men being regularly taunted with accusations that they "ought to be in uniform," which, according to Hart, were accompanied by "the added innuendo that they are dodging service in the Forces" (Hart, 1942b, p. 1). Such taunts were sometimes at the heart of many of the breach-of-peace charges against, and resulting court appearances of, unit members, following fights in local drinking establishments. As St. John Chadwick noted, such taunts had caused "quite a few broken heads and short prison sentences" (St. John Chadwick, 1942e, p. 1). Clement Attlee acknowledged that the NOFU's status as a civilian unit was one of "considerable difficulty" for many reasons, including exclusion from the extensive support services of the Navy, Army and Air Force Institutes (NAAFI) offered throughout the UK (Attlee, 1942a, p. 2).[49]

Some in the UK publicly defended the NOFU. The wife of one member wrote to Dundee's *The Sunday Post* to express her concerns that the men had come so far to support the war effort yet were "being rather scurvily treated" (Lumberjack's Wife, 1941, p. 5). Her specific examples highlight some of the indignities that came with un-uniformed service:

> One night last week a cinema show was given for the troops. Five or six lumberjacks walked four miles into the village, produced their badges to the soldier there, but were turned away. My husband, a lumberjack, had two friends in the Seaforths home on leave.[50] He accompanied them to the Forces canteen, and was admitted by a lady who knew me. Nothing was said to him. Yet he was made to feel he was not wanted because he had no uniform. He refuses to go back. (Lumberjack's Wife, 1941, p. 5)

---

49 The NAAFI operated a variety of catering and recreational services for the armed forces, including canteens. The Entertainments National Service Association, which later organized concerts and films for the NOFU, was an arm of the NAAFI.

50 The Seaforth Highlanders, an infantry regiment of the UK armed forces.

She concluded with the point that "if [the NOFU men] received more consideration, many of them would renew their contracts instead of going back home" (Lumberjack's Wife, 1941, p. 5).

But approving uniforms for the NOFU presented numerous difficulties, despite having support within various departments and offices. General Carrington backed the idea but feared access to uniforms might affect enrolment in the yet-to-be approved and long-delayed Home Guard unit, which was to be comprised entirely of men of the NOFU (St. John Chadwick, 1942e). As well, Timber Production Director Gerald Lenanton, never a fan of the NOFU men, indicated to C. S. Pyke of the Ministry of Supply that "the provision of uniforms [to the NOFU] would present great difficulties," given the demands of other units that would undoubtedly follow any concession to them (Lenanton, 1942a, p. 1). As evidence, Lenanton cited the uniform for the Women's Timber Corps formed a few months earlier as "one of the most difficult hedges to overcome" (Lenanton, 1942a, p. 1).

The Women's Timber Corps was eventually granted a uniform, but the NOFU was not. Despite efforts to differentiate the unit from other organizations based on its overseas status, the response of the UK Board of Trade, as conveyed by H. J. B. Lintott, was "most unfavourable" (Lintott, 1942, p. 1). Arguing that issuing uniforms to the NOFU "would seriously prejudice" the Board of Trade's handing of persistent requests for uniforms from other organizations, he wrote: "I doubt if it could be argued that the Newfoundlanders are in any very different position from many other workers who are playing a more direct part in the war effort without any uniform to advertise the fact" (Lintott, 1942, p. 2). Lintott encouraged Pyke to persuade "the people responsible" that issuance of uniforms would be "not merely inadvisable but improper" (Lintott, 1942, p. 2).

Emrys-Evans delivered the decision to Turner in a letter dated November 26, 1942, some two months after Hart had first presented the NOFU case to Carrington. He underscored Lintott's points, including emphasizing that NOFU workers did not meet the main criterion for uniforms: "the uniform is indispensable in order to enable the worker to do his job properly [and] if it is necessary for that purpose that he should be instantly recognizable and distinguishable from the general public" (Emrys-Evans, 1942f).

The NOFU men's desire for uniforms did not abate, however. Approximately half of the unit's strength joined the Home Guard and gladly wore its uniform, which differentiated them from their own comrades as much as it identified them as a home-front defence force.

A Combat-Ready (and Uniformed) Role: The Home Guard

Many of those who did not enrol in combat forces or who remained at home (in both Newfoundland and the UK and for whatever reason, including being ineligible to serve) nonetheless felt the urge to participate meaningfully in other efforts to support the war. One of these efforts was in home-front defence (as opposed to frontline combat). During wartime, the UK government turned to a volunteer army of men who did not have combat roles for this very purpose, as did the NCG.[51] First called the Local Defence Volunteers, the UK home army was formed in May 1940. Its duty was to provide interim defence in areas under threat until regular forces could be deployed. It was composed of men who were ineligible for combat service or who were in essential work occupations at home and therefore could not enlist.[52] In July 1940, Prime Minister Winston Churchill suggested its name be changed to Home Guard. By that time, over a million men had joined the home army (Summerfield & Peniston-Bird, 2000, 2007). They included men from the NOFU, who enlisted in units established in towns near their camps. In the early call for Local Defence Volunteers, preference was given to experienced firearms users; their main duty was to spot Axis parachuters and infiltrators.

---

51 The Home Guard in Newfoundland was established as voluntary defence units for the mill towns of Grand Falls and Corner Brook, which were considered especially vulnerable to attack. They were renamed the Newfoundland Militia in 1943. See Higgins (2007b).

52 While some women did work in support roles for Home Guard units, they were not legally members, despite efforts by advocates to make women full participants (including in preparation for combat). The Women's Home Guard Auxiliary was formed in 1943. See Summerfield & Peniston-Bird (2000) for a discussion of women's resistance to exclusion from the Home Guard, including the development of alternative organizations such as the Women's Home Defence.

The initial proposed Home Guard uniform was a light khaki denim overall and a forage cap ("New Defence Corps," 1942), which was worn until a more standard military-style uniform became available.

The NOFU men were enthusiastic volunteers in this home-based defence. When the call for volunteers was issued at a camp at Glenmuick, near Ballater, its sixty or so men enlisted en bloc. Larry Gladney joined a unit attached to the Royal Black Watch, with headquarters at Perth. Gladney was at Taymount camp, near Stanley, when recruiting officers from the Department of Defence visited. As identifiers, Gladney and his comrades were issued only an armband, an identification card, and a shotgun. There were no drills and very little instruction, except in how to make and use explosives. Later in the war, rifles and full dress uniforms, including a great coat, were issued to volunteers (Gladney, 1999).

The 3rd Inverness (Newfoundland) Battalion of the Home Guard

By the early fall of 1942, with the NOFU concentrated in the north of Scotland, another Home Guard unit was formed for the area that was drawn exclusively from NOFU members. Previously, approximately one hundred members had joined pre-existing units, but the distance between camps and towns, and the movement of men among camps, inhibited full participation. To address these obstacles, to give men an opportunity for military training, and to contribute to local security, the NOFU leadership requested permission to raise a battalion from the unit.

The request was finally granted in fall of 1942 (Turner, 1944, 1945b). Although it is unclear when the unit was first proposed, correspondence from Paul Emrys-Evans to Charles Peat dated October 14, 1942, indicates that the creation of the unit had been already "so long delayed" (Emrys-Evans, 1942e, p. 1). The uncertain status of the NOFU over its early overseas years was surely a factor in this delay. Once the crises of 1942 were resolved and the battalion subsequently formed, however, it boasted "the proud distinction of being the only Home Guard Unit in existence formed entirely of Dominion civilians working in this country on specialised war work" (NOFA, 1945, p. 57).

Officer-in-Charge Turner, assigned the rank of Lieutenant-Colonel in the Home Guard, was appointed Commanding Officer of the new unit, which was officially named the 3rd Inverness (Newfoundland) Battalion. Joseph Curran was Second-in-Command at the rank of Major; both appointments were effective as of September 30, 1942. James Gerard ("J. G.") Martin replaced Curran on September 27, 1943, following Curran's repatriation.[53] By the end of 1942, the unit included 750 volunteers; in 1943, through repatriation, battalion strength was reduced to 650. With the exception of a small few who had served in the First World War, the men had no prior military experience. Most were under the age of 45 (Turner, 1944).

The 3rd Inverness (Newfoundland) Battalion wore the uniform of the Home Guard, with the NOFU badge in the uniform cap and a Newfoundland shoulder flash that read "Newfoundland Home Guard 3 IN." It was initially designated as a mobile guard, and shared responsibility for certain locations with the 1st Battalion of the Inverness-Perth Road. It was eventually designated as a full mobile unit, given that the NOFU had its own transport, drivers, and mechanics, which enabled it to fulfill this role (Turner, 1944).

A company of the 3rd Inverness (Newfoundland) Battalion, Inverness-Shire, 1943.
(Courtesy of The Rooms Provincial Archives Division, A 61–45)

53 For a list of officers and other particulars of this Home Guard unit, see Turner (1943a) and NOFA (1945). Original training manuals, certificates, photographs, and other ephemera are held in the Maurice Job Taylor Fonds at The Rooms Provincial Archives, St. John's, NL.

The battalion was divided into four companies, one from each of the three districts and a fourth from Headquarters (A Company, B Company, C Company, and HQ Company). HQ Company was established in late 1943; the others were established on formation of the battalion (Turner, 1944). A, B, and C companies each had a commanding officer and a second-in-command, as well as platoon commanders and platoon officers. The HQ Company, based at NOFU operational headquarters at Carrbridge, had these same officers and, in addition, officers designated for intelligence, transport, signal (communications), weapons, and munitions (NOFU, 1945). Training was conducted after hours, on weekends, and during leave; officers and NCOs were trained on weekends so they could perform their own duties in training their platoons. Basic training was held in the winter of 1944 and additional training exercises followed. The men constructed their own assault courses and a rifle range at an abandoned camp in the foothills of the Cairngorms, near Carrbridge (T. V. Curran, 1987; Nicholson, 1969; NOFA, 1945; Turner, 1945b).

Unsurprisingly, in the context of war scarcities, a shortage of equipment plagued the Home Guard. Equipment was often donated by locals, but supply sometimes did not meet its needs. In some cases, an appeal to Newfoundland was made, such as in June 1943 when Turner requested binoculars of the Secretary of Natural Resources, K. J. Carter, who issued a public appeal with a promise that "binoculars will be properly cared for and returned unless lost through circumstances over which he [Turner] has no control" (K. J. Carter, 1943, p. 7). Given wartime economic prosperity at home, the NCG was in a position to provide some support for this request, but it did not.

In the fall of 1944, as an expression of respect for this unique unit of the Home Guard and in the tradition of other Inverness-shire battalions, members of the 3rd Inverness (Newfoundland) Battalion were named honorary members of the Regimental Association of the Queen's Own Cameron Highlanders ("Newfoundlanders Become Honorary Members," 1944). The unit was also recognized for its skills. In a competition of 1,080 Home Guard units from throughout the UK, including Northern Ireland, the Battalion placed third for the Lochboisdale Challenge Trophy of the National Rifle Association (Turner, 1945b). Turner provided context for this accomplishment in his public report of 1944, underscoring the extent of their success despite liabilities that included the men having little to no military background, their small and widely dispersed locations, and a training period of only two years compared to the four years the other battalions in the competition had trained (Turner, 1945b).

The 3rd Inverness (Newfoundland) Battalion operated for just over two years and stood down on December 31, 1944, after the threat of a German invasion had ended. At stand down, the "husky volunteers" were praised in the UK press for their initiative both in the Home Guard and for their work as members of the NOFU in the forests of Scotland ("Far-From-Home Guard," 1945, p. 7). As overseas British subjects, NOFU members who served in the Home Guard unit for a period of not less than six months were entitled to the Defence Medal upon proof of service in the 3rd Inverness (Newfoundland) Battalion—or any other UK Home Guard unit ("Public Notice," 1946). As well, several members received individual honours (T. V. Curran, 1987; Turner, 1945b).[54] Each member of the Home Guard also received a certificate of participation signed by HRH King George VI.[55] Representatives of the unit participated in the Victory Parade in London on June 8, 1945.

The 3rd Inverness (Newfoundland) Battalion enhanced the unit's reputation and increased goodwill between the NOFU and local residents. It also likely quelled the desire many in the unit held to serve in a role other than civilian worker. The unit formation came at a time when NOFU operations were becoming more centralized, yes—but was also at a time when contracts were more stringent about required time of service, and when transfers to other forces had become much more difficult. As well, the overall welfare of members was in the spotlight and discontentment in the unit was high (Peat,

---

54 John Nick Jeddore described how, in one instance, with so many worthy contenders for an award, there was a draw to determine the choice (Jeddore, 2001). For their leadership in both the NOFU and the 3rd Inverness (Newfoundland) Battalion of the Home Guard, in 1943 Jack Turner and Joseph Curran were appointed to the Most Excellent Order of the British Empire—Turner as an Officer (OBE) and Curran as a Member (MBE).

55 There is no official history written of this unit. What little is known is provided in NOFA (1945) and repeated later in Nicholson (1969) and T. V. Curran (1987). Its story remains an underexplored aspect of Newfoundland volunteer service in the Second World War.

1942b). Along with other measures to alleviate welfare concerns and enhance unit discipline, increased and convenient access to Home Guard activities was a positive intervention for many NOFU officers and members. The Home Guard unit's formation would also have signalled a degree of respect for the men of the NOFU by the UK authorities and gone some way toward alleviating tensions between the Dominions Office and the NCG.

Service in the Home Guard provided more than just training for potential combat service. It added a layer of legitimacy to the NOFU men, who had felt that their un-uniformed, civilian service was devalued both at home and in the UK. But this legitimacy would not stand the test of time because the Home Guard itself—its purpose and function—was contested during and after the war. Popular representations of it were often framed as humorous depictions that downplayed its contribution and ridiculed the combat-readiness of its volunteers (Summerfield & Peniston-Bird, 2007). In this sense, it was yet one more contested role that the NOFU men assumed—one that, yet again, did not fully align with the idealized man-as-combat-soldier that dominated wartime notions of masculinity and service.

At home after the war, Home Guard service became a dividing point within the unit. Enrolment in it was used as a criterion for postwar recognition, including membership in the Canadian Legion ("Canadian Legion Conference," 1960), thereby creating a two-tiered system of service recognition for ex-NOFU members as they struggled for privileges and benefits for all.

On Course for Citizenship and Change: Education and the British Council

The Home Guard was not the only avenue to additional training that the NOFU men could access during their time overseas. Other opportunities for training and education became available as a result of the welfare crisis of 1942 and as the question of the future of Newfoundland after the war came into greater focus.

When the Newfoundland Commission of Government was installed, the education system in the country was encumbered by challenges on several fronts. There were not enough schools and many of those in existence were dilapidated, teachers were underpaid and inadequately trained, and school fees and book costs were beyond the means of many families who, in the 1930s, were living in destitution. Where schools did exist, attendance rates were often low because the family subsistence economy prioritized home labour needs over education. As a consequence, children old enough to participate in family fishing or woods work were needed at home; child schooling was a luxury many parents felt they could not afford. Schools were governed by Christian denominations whose differences and competitive (rather than collaborative) positions often subverted the best educational interests of children and families. When the NCG was appointed in 1934, one of its first and most meaningful interventions was in education. It took measures to improve teacher salaries, build new schools, and address low enrolments through compulsory attendance legislated by the School Attendance Act of 1942 (McCann, 1987).

The men who enrolled in the NOFU were a product of this deficient educational context. Both educational and literacy levels in the country were very low, much lower than, for example, those of the neighbouring Canadian maritime provinces of Nova Scotia, Prince Edward Island, and New Brunswick (Alexander, 1980). Low literacy contributed to merchant class dominance over a controlled and dependent population of fisher families who worked for them (Curtis, 1990). Limited literacy resulted in constraints, including the ability to express one's grievances, challenge servitude, and communicate generally. These limits were noted by many of those responsible for the NOFU's needs while the men were in the UK.[56]

In the woods camps at home, men with low literacy relied on those around them who could read and write to record their dictated letters home and read aloud those they received.[57] This practice continued in the NOFU camps. There were also volunteers from nearby communities who encouraged

56 For a comparable discussion of literacy and the men of the Newfoundland Forestry Companies of the First World War, see Kelly (2020).

57 For a discussion of camp literacy practices, see Kelly (2018). See also Barrett (1940), who wrote of reading aloud to NOFU members while travelling to the UK aboard the SS *Chrobry*.

literacy development among the men. For example, Betty Lavers Bode, a nurse who joined the army during the war and who, while training at Strathpeffer, spent her spare time writing letters for some of the NOFU men who worked nearby ("Obituaries: Mrs. B. L. Bode," 1997). There were also dedicated teachers in towns and villages near NOFU camps who offered their expertise to the men in the evenings, as part of an informal educational effort to build literacy and numeracy among members of the unit.

Commissioner of Natural Resources J. H. Gorvin wrote to the Secretary of State for Dominion Affairs as early as January 1940, just after the third NOFU contingent arrived in the UK, to request educational support for some of the men who, upon return, could then participate meaningfully in the reconstruction work of the NCG. Gorvin suggested that certain "picked men" could be given

> some insight into grading and organisation, and technical operations such as black smithery and care of cutting tools, in commercial sawn lumber and pit prop industries in Great Britain; also reafforestation work of Forestry Commission and technical educational activities by Ministries of Education and Labour. (Gorvin, 1940a, p. 1)

For "two or three of [the] more intelligent," Gorvin suggested an apprenticeship in High Wycombe furniture-making or other wood-crafting industries that might arise in the country after the war (Gorvin, 1940a, p. 1). He added: "Any general elementary educational instruction for the greater number of the loggers during the period they are in Great Britain . . . would facilitate our later reconstruction work" (Gorvin, 1940a, p. 1). Gorvin's suggestions reflected his educational beliefs about the kinds of instruction best suited to the men of Newfoundland, including those of the NOFU (McCann, 1987; Neary, 2012) and fell in line with the NCG's reconstruction plan.

The suggestion of educational training for the NOFU men was initially nixed by Sir Roy Robinson of the Forestry Commission, who felt that "all energies must be concentrated on production, since the [UK timber supply] position was precarious" (Dominions Office, 1940, p. 2). Robinson's reply to the suggestion of education opportunities for the NOFU men was consistent with his views on the workers, whose value, he felt, lay only in their physical abilities. They were the sum of their labour and nothing more.

Others held a wider view of the potential of the men and what their well-being entailed. In his June 1942 Ministry of Supply report responding to welfare conditions in the NOFU camps, Charles Peat included a caveat about the limits of any welfare measures, given other problems that required attention:

> A high proportion of the men, I am told, are illiterate, many of them being boys under twenty. I feel that their time in this country, particularly in the long winter evenings, has been wasted and that an effort should be made to get them to make use of their opportunities to educate themselves and even fit themselves to take up occupations other than lumbering when the war is over, as I imagine that the openings which may be ready for them when they return to Newfoundland may be few and far between. (Peat, 1942b, p. 2)

Peat's comments, delivered a couple of years after Gorvin's request, aligned with the reconstruction objectives of the NCG. They also portend the collapse of the wartime economy in which the men would find themselves. Whatever the future, an educated citizenry was essential to any successful post-NCG governance.

Press accounts about the NOFU men sometimes touched on matters of education and learning. As one writer observed: "These stocky, muscular men from the fishing grounds and forests have an engaging simplicity and sincerity, and, if some of them 'ain't got no learnin'—as a few of them immediately confess—they have a sharp intelligence" (Ley Smith, 1941, p. 4.). The building of education huts in NOFU camps as part of the 1942 overall welfare recommendations addressed some of the educational needs of the men that had not been met at home. Huts were annexed off canteens and "Camp School" was offered in them from 1943 onward (NOFA, 1945). St. John Chadwick of the Newfoundland Branch of the Dominions Office reported early in the summer of 1944 that:

> With the assistance of the British Council and a number of energetic local school teachers who have generously given up their free evenings to this work, a man can learn anything from simple grammar to advanced mathematics, and several loggers who could not sign their name when they arrived in this country are now cheerfully writing letters home to their wives and fiancées. (St. John Chadwick, 1944, p. 5)

One such teacher, D. R. Morrison, science master at Kingussie School, was the first to volunteer. He was featured on *Scottish Half-Hour*, a BBC radio program offering local content, speaking about his educational work with the NOFU men ("Scottish Half-Hour," 1944).[58] Following a visit to the NOFU camps in the spring of 1944, Sir William Brass also commented on the work of Mr. Morrison, whom he noted had "done particularly good work at great personal self-sacrifice" (Brass, 1944, p. 1). Morrison represents the dozens of teachers whose efforts on behalf of the NOFU men were part of their support of the war effort.

In 1944, Brass was also able to report to Paul Emrys-Evans that "a very happy state of affairs" existed in the camps—a marked improvement from his visit during the crisis of 1942 (Brass, 1944, p. 2). Commenting on the provision of education huts, Brass noted that "the Newfoundlanders welcomed this innovation and made considerable use of these opportunities for study" (Brass, 1944, p. 1). As part of the educational initiatives for the NOFU, correspondence courses in a variety of subjects were also made available, as were vocational education courses in areas such as welding, auto mechanics and carpentry, which were offered at Inverness and Aberdeen on Saturdays during the men's time off (Turner, 1945b).

After the war, educational opportunities in trades such as those offered to the NOFU were made available in Newfoundland through the establishment

---

58 Morrison later wrote a history of Kingussie School and was provost—or mayor—of Kingussie for several years.

of the Vocational Institute for Ex-Servicemen. Alfred E. Hart, the Administrative Officer for the NOFU and a senior employee with Timber Production, was hired by the NCG as the founding director of the institute, a division of the Education department that eventually targeted civilians and marked the successful inauguration of vocational education in Newfoundland ("Trade School," 1947).[59]

The British Council

The educational efforts on behalf of the NOFU were overseen by the British Council. Formed in 1934 as the British Committee for Relations with Other Countries, the purpose of the British Council (as it became known shortly thereafter) was to promote knowledge of British culture and language abroad and build ties between the UK and other countries. A response to increasing Axis propaganda in the 1930s and the decline of UK influence abroad, the fledgling organization found its role transformed in the war. By 1940, the Home Division of the Council became responsible for the cultural and educational well-being of the thousands of Allied civilians living in the UK during the war (Fisher, 2009).

The Aberdeen office of the British Council, in whose jurisdiction the NOFU camps fell, had offered courses for several Allied foreign nationals in Scotland, including Canadians. By 1942, in response to concerns raised about the educational and literacy levels of NOFU members, and coincident with an emerging clarity about the imminence of a change of governance in Newfoundland, the Council had established educational programs for the NOFU camps. It arranged correspondence courses, distributed books, newspapers, and magazines to camps, and promoted other educational and cultural activities such as visits to industrial and research centres (Bruce, 1945).

Along with general literacy and numeracy recommendations, in 1942 Charles Peat also suggested that "it would be a good thing if some of [the NOFU men] were given an opportunity to visit London and Edinburgh and

59 In recognition of his work with the NOFU, Hart was appointed a Member of the Most Excellent Order of the British Empire (MBE) in 1946 ("Awarded MBE," 1946). He immigrated to Newfoundland after the war and stayed until 1950.

to have an opportunity of learning something about the traditions and history of the United Kingdom" (Peat, 1942b, p. 2). In this regard, the Council also arranged five-day "special leave" courses. These included lectures and visits to industrial and cultural sites. For example, a course on the Scottish fishing industry featured presentations, inspections of trawlers, tours of dehydration, curing, and refrigeration plants, and visits to fish markets and the Torry Research Station—and included stops at the historic and cultural sites of Aberdeen ("Nfld. Forestry Unit Attends Study Courses," 1944).

The UK government and the British Council felt some responsibility to prepare NOFU men for life at home after governance by the NCG—a wartime educational investment in preparation for future divestment of its responsibilities toward the colony. At one of the several courses in fishing offered at Aberdeen, Robert Bruce, who in 1943 was appointed to head the Aberdeen office of the British Council, put the goal of the studies in which the thirty NOFU men in attendance were engaged into this context: "It is the wish of all of us that full self-government [in Newfoundland] will be achieved in the near future and that such courses as these will help in giving Newfoundland an opportunity of studying our methods of industry and government" in preparation for that future ("Old World," 1944, p. 4). To focus educational pursuits, the Newfoundland Overseas Foresters' Association (NOFA), a new member organization for men of the NOFU formed in 1944, distributed a questionnaire and created sub-committees for resulting areas of interest; they included farming, fisheries, local government, timber, and unemployment and health insurance ("Sub-Committees," 1945).

Courses explaining the UK's system of parliamentary government, such as the one offered by the British Council in spring 1946 at the University of Aberdeen's Marischal College, were part of an effort to educate Newfoundlanders and Labradorians about democratic governance. The twenty-five attendees all held "executive positions" (camp and district leaders) in the NOFU ("Newfoundlanders at University," 1946, p. 3), perhaps reflecting a hierarchy of interest between the officers and general membership of the NOFU. One-hour lectures in local governance were also offered at the district level at camps at Carrbridge, Revack, Laggan, and Kinveachy in February 1944, in each case with between sixty and ninety men in attendance (A. L. Stewart, 1944). Three-day courses on local governance were also offered at Inverness, with dozens of NOFU members participating ("Six Towns," 1946).

Meanwhile in Newfoundland, the NCG had created the Division of Local Government within the Department of Health and Welfare (1943). Public information was distributed to create interest among local residents about the services available through incorporation ("Six Towns," 1946). Several towns on the Island were incorporated during the war, including Corner Brook, Grand Bank, Channel–Port aux Basques, Lewisporte, Wesleyville, St. Anthony, and Harbour Grace, as well as the districts of Placentia and Springdale–South Brook. The education of NOFU members on issues of local governance dovetailed with these efforts to build a local governance structure for the country by enhancing leadership capacity and supporting it through focussed study for the men of the unit.

When the future of Newfoundland was debated in the UK House of Commons, Under-Secretary of State for Dominion Affairs Paul Emrys-Evans, responding to questions from Conservative Member Beverley Baxter, commented on the importance of education in local governance: "It is felt that the political education of Newfoundland can best be achieved by encouraging local government development." He added that "steps have been taken to provide political education for Newfoundlanders serving in the forces and in the Newfoundland Forestry Unit" (Appel, 1945, p. 7). As Peter Neary noted in commentary on the evolution of democracy in Newfoundland, the late arrival of local governance structures—"an essential training ground to democratic practice" (Neary, 1969, p. 42)—was one factor in the delayed growth of democratic processes in the country.

Whether a political education of Newfoundlanders and Labradorians was needed was also debated. Communist Member William Gallacher wondered if the notion implied that they were "a backward race or incapable of looking after their own affairs" (Gallacher, 1945, col. 14). A news political correspondent who heard the debate in the House of Commons was recorded as saying, "the Newfoundlanders I've met over here didn't seem to need any political education" (Wigham, 1945, p. 3). One parliamentary correspondent urged authorities in Newfoundland to be "more publicly conscious," and to ensure that the press has the information it needs "for telling the world about the valour of your sailors and the tenacity of your soldiers; and about the tremendous work that is being done by your Forestry unit" (Wigham, 1945, p. 3).

These studies were thus both enriching for the men of the NOFU and part of an overall effort by the British Council to influence the development of citizens of the dominion, in keeping with the Council's purpose (Eastment, 1982). Courses also fitted into the reconstruction plans of the NCG, as well as the UK government's desire for the re-establishment of some form of democratic governance in Newfoundland, which would allow the UK to step back from its responsibilities in a postwar era that was sure to bring too many demands on a public purse emptied by the costs of that conflict.

The NOFU men were also eager to embrace opportunities to exercise their rights, not only as educated workers, but as subjects interested in shaping the future of their country. Through letters and newspapers, they kept abreast of happenings at home—summaries of news items were published in the *Newfoundland Foresters' Bulletin* (herafter, the *Bulletin*), the official publication of the NOFA—and they viewed their studies not only in terms of personal development but also for the potential impact of this knowledge at home. In one issue, the editors optimistically expressed a common refrain from previous issues, that the unit, "with our accumulative knowledge gained from mixing with people in this country, attending lectures and courses, should prove to be a strong factor in any decision which may govern the future of our native land" ("Editorial," 1946, p. 1).

When Governor Gordon MacDonald visited the unit in 1946, as the final group of NOFU workers was preparing to leave the UK, he observed this deeply felt commitment. In a letter to the unit, he commented on "the keen interest shown in the future of Newfoundland" and "the clear conception shown by men of the problems and also of the possibilities in their homeland" (MacDonald, 1946, p. 1).

From Desperation to Dreams: Organizing for Newfoundland's Future

When the NOFU workers departed for the UK in 1939, they had lived for five years under rule by the NCG. They left behind a country characterized by destitution and struggle, one in which the people had no say in government affairs that affected their daily lives. The NOFU members' impressions of the NCG were in keeping with those of their compatriots—largely negative—and many advocated loudly for a return to self-government. By war's end, Newfoundlanders and Labradorians had been without self-government for over a decade. The younger men of the NOFU had never experienced the franchise or adult life in a democracy.

In an invited submission for the *Bulletin* in 1945, subsequently reprinted in *The Daily News* of St. John's, Robert Bruce of the British Council's Aberdeen Office wrote about his impressions of Newfoundland that, according to him, were shared by many in the UK.[60] They included surprise that Newfoundland was, at the time, "not a self-governing Dominion in the full sense" and shock that education facilities were poor, especially in outports, adding that despite full knowledge of the conditions that prevailed or the reasons, many concluded that "something is very far wrong in Newfoundland" (Bruce, 1945, p. 7).

Bruce commented directly on the men of the NOFU, in a manner resembling some of the news coverage that had greeted their arrival in the UK. He saw in the NOFU:

> [A] group of men who are straight in their dealings, hardworking, willing to take risks ... whose outlook is natural and spontaneous, that of the woodsman and the fisherman. I ... sense an atmosphere of shy friendliness and easy-going equality among everyone from superintendents to loggers. You give one an impression of being far away from cities and office desks. (Bruce, 1945, p. 7)

60 Newspaper editors in Newfoundland regularly reprinted articles of interest from the monthly publication of the *Bulletin* or used some of the opinions expressed therein as the basis for an editorial. See, for example, an editorial on local manufacturing in *The Evening Telegram* ("Obstacles," 1945).

Bruce acknowledged the work ethic of the men and their productivity, but he also noted that there was a quality—citizenship—that was missing among them, identifying the source in which he felt "some of Newfoundland's difficulties can be found" (Bruce, 1945, p. 7).

While the NOFU workers considered their country from afar, as it edged toward decision-making about its future in the mid-1940s, Bruce's point likely resonated. Developing citizenship was at the heart of the educational efforts of the British Council as they related to the NOFU and to the emerging question of the future of Newfoundland, but addressing the "political immaturity" related to its absence was also an obstacle to a more complete democracy. The roots of this issue reached well into the country's past, making addressing it a larger project that would extend far beyond the war years and rule by the NCG, and well into the era of Confederation with Canada (Neary, 1969, p. 44).

The Goodwill Mission

As part of further information-gathering and pulse-taking in relation to Newfoundland's future, Clement Attlee, now Deputy Prime Minister and Secretary of State for Dominion Affairs, appointed a parliamentary committee in 1943. Its "Goodwill Mission" to Newfoundland was an opportunity to hear directly from the people their views on the country's political future. The Mission was led by Labour Member Lord Charles Ammon and included two other parliamentarians: Conservative Sir Derrick Gunston and Independent Alan P. Herbert. The three travelled throughout the Island and Labrador for three months in the summer of 1943 ("British Parliamentarians," 1943; "Record,"1944). Their work was informal and no report was required.

Nevertheless, in 1944 Lord Ammon published his perspective on what he had observed in a booklet entitled *Newfoundland, the Forgotten Island*. Ammon argued that, based on his observations, the people of Newfoundland and Labrador were neither prepared for nor wanted a return to self-government. He recommended, instead, a transition marked by a modification of the NCG to be more representative—through election of the Newfoundland Commissioners—and the creation of a ten-year development plan for the country under its rule. Ammon concluded that "to hasten slowly is the most likely method to secure a safe arrival" to stable self-government and secure social development (Ammon, 1944, p. 11). His publication was at the centre of a tense exchange in the House of Commons on December 2, 1943, during which Paul Emrys-Evans fielded questions about the political education of Newfoundlanders and Labradorians and the status of plans for self-government for the country from several Members of Parliament, including Conservative Beverley Baxter, a vocal supporter of self-government, and Independent Alan Herbert (Wigham, 1945).

Out of those exchanges arose a commitment to debate the Newfoundland question two weeks later, on December 16. During a five-day visit to the NOFU camps prior to that debate, Sir Derrick Gunston gathered the men's perspectives on their homeland (St. John Chadwick, 1944). Impressed by what he heard and observed, Gunston commented during the House of Commons debate that

> by an extraordinarily fortunate coincidence we have in this country at this moment the Newfoundland lumber unit . . . [which] is doing a great work in Scotland. . . . I suggest that this unit ought to be kept together—they are helping the United Nations to win the war—[and] after the war let them go back to Newfoundland as a unit and help to develop their own country. (Gunston, 1943)

Some NOFU members shared Gunston's thinking. Two years later, as they prepared for repatriation, the executive of Deisher Camp posed the possibility of the unit remaining intact in postwar Newfoundland "to work in some timber area on a co-operative basis for the export of timber" and to provide a settlement for interested members and families ("Camp Meeting," 1945, p. 9). The comments spoke to the social thinking, camaraderie, and sense of shared purpose that had evolved, at least among some in the unit, during the men's time away from home.

Thus, in the later years of the war, this large group of overseas Newfoundlanders and Labradorians caught the attention of parliamentarians in new ways, as both a target for political education and, through it, an instrument of social reconstruction. Several months after Gunston's visit, the NOFU itself created the organizational means by which its members would maintain affiliation after repatriation, for what it saw as the benefit of its members and their country. By this time the House of Commons had passed a resolution endorsing Newfoundland's right to self-government. The future, while still unclear, was nevertheless unfolding, and many of the NOFU men were preparing to be a productive part of it.[61]

The Newfoundland Overseas Foresters' Association

Many of the men of the NOFU clearly saw both a responsibility and an opportunity to maintain ties with one another and to influence collectively the direction of decision-making in their homeland. To these ends, individual camp meetings were held during the summer of 1944 (Turner 1945b). A general meeting followed at Fairburn Camp on August 20, 1944, at which the basis of the constitution for their organization—the Newfoundland Overseas Foresters' Association (NOFA)—was formed through member agreement ("Notes," 1944). The new organization's constitution was published on October 1, 1944 (NOFA, 1945).

The NOFA constitution outlined the democratic, non-sectarian, and non-affiliated nature of the organization, as well as its commitment to preserve and extend the comradeship that had become a hallmark of the unit. It sought to ensure just treatment of members on their return home, and to encourage their capacity to "take their full share in the work of making Newfoundland a country in which they and all other citizens may lead a good life" (NOFA, 1945, p. 68). Its allegiance was clear: "The Association exists to work for the welfare of Newfoundland and the Empire" (NOFA, 1945, p. 68).[62] In its attention to social unity, improvement of education, reorganization of local and national government, and a spirit of co-operation to work with other organizations "whose objects are the development, welfare and improvement

---

61 For an account of this and other developments related to the future of Newfoundland during this period, see M. Baker (2003).

62 For the full Constitution, see NOFA (1945).

of conditions in Newfoundland," the NOFA echoed many of the aims and objectives of the British Council's engagement with the NOFU.

The *Newfoundland Foresters' Bulletin,* which was published monthly between November 1944 and April 1946, was the official organ of the NOFA. The *Bulletin* was a combination opinion platform, information flyer, and social column. Its regular features included: an editorial; letters to the editor; NOFU news ("About Ourselves"); Newfoundland news ("News From Home"); camp and committee reports; contributions from members; reprints of lectures, short articles, or correspondence; news of hospitalizations, deaths, births, and marriages; a list of visitors to the camps; and other news items (such as lists of member repatriations, etc.). Issues typically ran to four single-spaced, two-column pages and were printed at Beananach, Carrbridge.[63]

The *Bulletin* quickly became a sounding board for opinions about the future of Newfoundland, the role of NOFU members in shaping it, and the reintegration of NOFU members upon repatriation. It was a place for important discussion of issues related to democracy, citizenship, and related rights and responsibilities. The *Bulletin* was sent to Newfoundland newspapers, who published bits of content from it, ensuring that there was some information circulating to the public at home that represented the voices of the NOFU men themselves speaking about not only their work but their ideas about the future of the country.

In 1945, early in the NOFA's formation, its executive also compiled and published a record of the unit's time in the UK. It was designed "to illustrate some of [the] work it has done since the unit left Newfoundland, to tell the story of its life, and to make available a souvenir" ("Editorial," 1945a, p. 1). Entitled *Timber!! Newfoundland Overseas Forestry Unit,* it was the first published and pictorial account by the unit of its formation and work, and it contributed to some family and public knowledge of the unit. It was sold to members and the general public with profits helping fund the work of the NOFA on behalf of its members.

---

63 While a complete set of every issue of the *Bulletin* is not available, some issues can be found at The Rooms Provincial Archives, St. John's, NL, and the Aberdeen City and Aberdeenshire Archives.

The educational and organizational experiences of their final years in Scotland enhanced the political thinking of NOFU members. The NOFA and the *Bulletin* gave workers forums through which to express and exchange ideas—to share concerns about the homeland and rule by the NCG that many members held prior to living there again.

A few months after his arrival in the UK in 1940, Temple Butler (NOFU #466), a farmer from St. George's stationed at Kielder, Northumberland, wrote in a letter home about the collective sense among the men in his camp of the situation in Newfoundland:

> We get occasional papers from home and devour them. One notices . . . that political life in Newfoundland is by no means dead. It is heartening, though we think the present is not the time for a reversal of government form in Newfoundland. We are proud that the democracy that we are fighting for is being kept in reach in the home country. We lumberjacks are playing a vital part in the struggle and Newfoundland may well be proud of the efforts of her sons. (T. Butler, 1940, p. 5)

His comment, written long before the later efforts to focus attention on these matters, challenges the contention that Newfoundlanders and Labradorians were not thinking about the political context in which they lived, or did not want a return to self-government. In 1943, Butler published a pamphlet about the future of Newfoundland. *Blueprint for Newfoundland: An Industrial Prospectus for Newfoundland's Development* was written while he was in the UK. It was initially sold in Newfoundland and eventually distributed free-of-charge. His writing caught the attention of the press and parliamentarians ("London Letter," 1944). James Maxton, an Independent Labour Party Member who met with Butler at the House of Commons on February 10, 1944, urged all parliamentarians to read Butler's ideas, proclaiming to the House that Butler "was as intelligent a man as I have ever met" and that his ideas were "as sound as anything produced by our three Commissioners" (Maxton, 1944, col. 2021).

In addition to Temple Butler, many other NOFU members shared their ideas through a variety of forums, including NOFA camp committee discussions and letters to the editor of the *Bulletin*. In this way, the organization fostered debate, encouraged disciplined thinking, and prepared members for fuller participation as citizens of a democracy that so many of them craved. The NOFA challenged the image conveyed by the *Amulree Report*—that Newfoundland was populated by a "simple-minded electorate" (NRC *Report*, 1933, p. 82)—showing them, instead, to be thoughtful, collective-minded workers who were intellectually engaged and emotionally invested in the future of their country. Robert Bruce praised the NOFA for its display of "social thinking" and sense of "communal responsibility which is essential to social progress," in what he regarded as "a stride forward in the development of citizenship in Newfoundland" (Bruce, 1945, p. 7). Its focus on the general welfare of Newfoundland through education, industry, and governance represented "the first steps towards a more complete democracy" (Bruce, 1945, p. 7).

The small dues required for membership in the NOFA generated some income for the organization. Members were given a pocket handbook that included the constitution, and a membership card. The organization also sponsored fundraising events in the districts, usually dances and card games such as whist.[64] With the exception of a small working balance, funds raised were invested in War Savings Certificates for use in the fulfillment of NOFA goals ("Notes," 1944).

Initially, the executive of the NOFA hoped to enrol most of the over three thousand men who had been part of the NOFU. One year into its formation, there were over six hundred members. Despite efforts to increase enrolment through camp drives and outreach, that number did not rise significantly. By March 1945, the NOFA had 667 members ("About Ourselves," 1945), and by the fall of that year was at its wartime peak of just over seven hundred ("Foresters Face," 1945), a number that remained stable as members repatriated. In the Highlands, the men faced an obstacle to fuller participation that was well-known at home—the impact of geography and

64 The available issues of the *Bulletin* contain regular references to camp dances and other events, the proceeds from which went to the organization.

the widespread dispersal of the population. Although the NOFU was concentrated in the north by the mid-1940s, some of the camps were far from the centre of operations at Carrbridge and this distance likely impacted membership numbers.

The significance of the NOFA was appreciated by many at home. The editor of *The Western Star* marked the first anniversary of the organization, noting hopefully that "a large well-organized association whose members are conscious of the fact that only the combined efforts of individuals will ensure successful development of the country, and a higher standard of living, is an asset to any country" ("Foresters Face," 1945, p. 2). But once at home, NOFA members were not only widely dispersed but also collectively derailed both by the lack of support for their postwar integration and the political groundswell of support for Confederation with Canada. As was the case when they arrived in the UK, upon repatriation the men found themselves still with "quite a lot to learn" (Turner, 1943a, p. 9). The importance of the Association, however, would emerge over the postwar decades and contribute to its goals—the well-being of its members and the betterment of its homeland—in ways that had not been imagined while its members were in the UK.

Other After-Work Preoccupations: Fundraising, Leave, and Visitors from Home

Involvement in the Home Guard and the NOFA, along with the courses of study in which many in the unit were involved, were clearly efforts by the men to become better citizens and to contribute not only to their own well-being and development but also to the communities around them in ways other than through their war work. While notions of citizenship among the men of the NOFU may have seemed to some as embryonic at first, their acts of benevolence and charity were fully formed—something the men demonstrated when, for example, they erected at their own expense grave memorials for NOFU members who died while in the UK. As well, recreational activities organized by the welfare officers of each district included, for example, sports competitions and dances that usually had a designated fundraising component for causes in both the UK and Newfoundland.

War Certificates

The war itself was the preoccupying beneficiary. When, with the introduction of piecework, members of the unit began to earn more cash, many men wanted to invest their surplus funds in war savings certificates. They initially purchased UK certificates under the War Savings Campaign, but they encountered difficulties cashing them at home. While in Newfoundland on recruitment work, Joseph Curran reported to Commissioner of Finance Ira Wild that some of the men considered the UK certificates as worthless. Curran cited examples of men purchasing certificates to support a local War Week or fund and then simply destroying them. He further explained that, without good opportunities to save, some men had squandered their extra funds on "extravagant expenditures, e.g., second-hand motor cars and drink" (Wild, 1942, p. 2). Wild followed up with Hugh Tait of the Dominions Office to set in motion the process by which the men could purchase certificates issued by the NCG. Paraphrasing Curran, whose points he must have felt would resonate in the UK, Wild wrote to Tait that "quite apart from the importance of saving ... [purchases] should curtail extravagant spending and encourage sobriety, which might very well result in increased output" (Wild, 1942, p. 3).

The War Savings Certificates program of the NCG was a registered savings plan begun in June 1940. It was designed to encourage Newfoundlanders and Labradorians to save while also supporting the war effort. Holders earned 3 percent compound interest over a six-year period on a certificate purchased at a face value of $2.50—a total earnings of fifty cents—with interest paid to purchasers by the government. A maximum of two hundred certificates could be purchased in a one-year period. A certificate could be redeemed at any time after six months from purchase date. Funds accrued through the War Savings Certificates program were used to meet interest charges on the public debt of Newfoundland and help the UK with war funding (*War Savings Certificates*, 1940).

The NOFU men were enthusiastic participants in the War Savings Certificates program, showing their commitment to the war effort and surpassing sales targets during any given drive. Through the purchase of War Savings Certificates to support the Wings for Victory targeted savings campaign, the NOFU contributed to the purchase of war planes. During the Greater London Wings for Victory drive of March 1943, for example, the men purchased a total of over $20,000 in certificates, more than doubling the initial target for sales ("Loggers Invest Savings," 1943). In July that same year, the unit purchased over $23,000 in certificates, for an overall unit total for the year-to-date of $90,000 ("Forestry Unit Purchases," 1943). One NOFU official noted that "some invested the savings of three years—every cent they had laid aside since they left Newfoundland" ("Foresters Surprise," 1943, p. 4).

The War Savings Certificates program ended in March 1945. Over $3 million in certificates had been purchased. Because their work time overseas had been extended, NOFU members were eligible to purchase beyond this date, suggesting the interest of the NCG in maintaining the savings practices of the men for the rainy day of postwar reintegration. Half of the total raised through the program was provided to the UK government as interest-free loans. Of the more that 55,000 Newfoundlanders and Labradorians who participated in the program, officials lauded the responses of school children and the men of the NOFU as particularly gratifying (A. M. Fraser, 2010, p. 118).

Other Fundraising

The generosity of the men extended to the local UK communities of which they were a part. In 1943, in a particularly poignant gesture, District C of the NOFU established the Olive May (Sybil or "Sybie") Clarke annual prize in Physical Education at Kingussie Secondary School, Inverness-shire. The prize, generated through interest on Defence Bonds donated to the school by the NOFU, was to be awarded in perpetuity and commemorated a "cheerful and efficient" employee and former student of the school (Brewis, 1943, p. 2).[65] On June 17, 1942, Sybil Clarke, 20, was killed in a lorry accident at the Kingussie Railyard, from where NOFU timber was shipped and where she worked

65 Defence Bonds were one of the savings groups within the UK War Savings Campaign during the Second World War.

("Newfoundlanders Win Esteem," 1943). The men were no doubt also aware that only a year or so earlier, Constable Fred H. Clarke of the Police War Reserve—her brother—was killed in a bombing in Glasgow (Strang, 2023).

The NOFU also participated in or sponsored fundraising for other local causes through events that featured the sports talents of its members. Boxing had grown in popularity as a pastime. A tournament at Ballater in April 1941 featured men of the NOFU and the British Army's Number 3 Commando; a collection was taken to support the Scottish Red Cross (*Fighting for the Red Cross*, 1941). The men displayed their boxing talents to Deputy Prime Minister Clement Attlee when he visited some of their camps in early September 1942, winning them headlines in the local press ("Mr. Attlee with Lumbermen," 1942) along with a featured photograph of camp boxers in action ("Lumberjacks at Play," 1942, p. 5).

The men also used their musical talents to support fundraising for various causes. The Scottish Youth Hostels Association noted this penchant for entertainment and its potential when announcing new memberships from the Glenfinart Camp, Argyllshire, and the expectation of more to come:

> Several of the Colonials state they intend seeing Scotland by the hostels while they are residing in this country. The lumberjacks will make excellent hostel entertainers. There are some splendid artists among them, and recently they gave a successful hillbilly concert in Dunoon. It has been noted that some of them might be an asset at the Glasgow District S.Y.H.A. show to attract new members. ("Youth Hostel," 1940, p. 15)

The musical talents and bunkhouse entertainment in the NOFU camps had been regularly covered by the press from the time of the unit's arrival in the UK. Published descriptions of bunkhouse times—a colloquialism for a gathering that included music, singing, and dancing—used subheadings such as "Loggers Like a Song" ("Lumbermen Like," 1941, p. 6) and included comments about the men's fondness for "dancing and amusements" ("Entertainment," 1940, p. 11). While local community organizations responded with events of their own to entertain the men, in later years welfare officers in each district organized events that featured both NOFU men and local performers. Chief Welfare Officer H. G. Penfold, an accordionist, often performed along with the men. Proceeds were donated to various charities including Barnardo's Homes for the care of disadvantaged children and Welcome Home, which welcomed returning service personnel.

Nor was the generosity of the unit confined to efforts abroad. A standout example of unit generosity was a $2,123 donation to the Notre Dame Bay Memorial Hospital restoration fund ("Forestry Unit Sends Donation," 1943). The hospital, which had been built to commemorate those who died in service in the First World War, opened in 1924 at Twillingate and was extensively damaged by fire on February 28, 1943. The NOFU included hundreds of workers from communities in Notre Dame Bay whose families were affected by the fire, and the men were especially moved to send what they could.

Leave, Travel Privileges, and Visitors from Home

NOFU men were allowed a half-day on Saturdays and all day on Sundays for personal pursuits, but there was little opportunity for travel. Although other units, for example, the British Honduras Forestry Unit, received special privileges for free travel, that was not initially the case for the NOFU. It was not until October 1943—almost four years from the first contingent's arrival in the UK—that free rail trips throughout Scotland were also granted NOFU men who wished to avail of unpaid leave to travel. As was shown in other instances, such as conditions around the use of petrol for recreation purposes, a paternalistic attitude toward the Newfoundland colonials was interwoven into most treatment of the NOFU and this privilege was no exception—to receive a pass, a member was required to have an absence-free work record from the time of the program implementation, and status as a good worker (Acton, 1943).

Although it was not easy for members to get leave and travel beyond the vicinity of the camps, the forestry camps were a popular destination for Newfoundlanders and Labradorians serving in other forces, as they had been during the First World War. When on leave or recuperating from injury, many soldiers and sailors headed to the Highlands. As one newspaper correspondent wrote: "It's a common sight to see a serviceman with a Newfoundland flash among the woodsmen. They look upon the forestry camps as a little bit of Newfoundland in Scotland" ("Far-from-Home Guard," 1945, p. 7). Often, these visitors were relatives and friends of NOFU men; some were ex-NOFU members themselves, returning to visit comrades or family. After the NOFA was formed, the *Bulletin*, in the fashion of the newspaper social columns at home, carried announcements of such visitors.

As had occurred during the First World War, a special bond formed between the Newfoundlanders and Labradorians and the Scottish people. Ex-NOFU member and gunner Wallace Butt (NOFU #2472) of Petries explained that while "the English people are very hospitable to us and they do lots to make things pleasant for us here . . . we are even more at home among [the Scottish people]" (Reyburn, 1942, p. 1). Butt added: "And, of course, we have the chance of getting together with a lot of pals from back home who are up there in the forestry units" (Reyburn, 1942, p. 1).

Reporting on one such visitor to a camp in Aberdeenshire, a journalist waxed sentimentally about the steady stream of ex-NOFU visitors: "Whenever they obtain leave they come back to the camps to see their old friends and hear again the pleasant accents of the Newfoundland forests and the squid-jigging grounds" (Ley Smith, 1941, p. 4). Along with the opportunity to reunite with relatives and friends, such visits offered the NOFU men a chance to share their enthusiasm for their new home. Victor Butler (#211413, Newfoundland Navy) of Port Rexton visited his brother Edgar (Ted) Butler (NOFU #2684) at a camp near Ballater while he was on a five-day leave in the spring of 1941. He wrote to family about travelling to a nearby camp, seeing historic landmarks, enjoying the scenery, attending church, and visiting the grave of James Butler (NOFU #1692), also of Port Rexton (V. L. Butler, 1941).[66]

There was another destination that the men sought out as a reminder of home. In an article about overseas Newfoundlanders and Labradorians in all

66 Butler had died the previous summer and was interred at Tullich Old Churchyard, near Ballater.

war service sectors and the role of the War Comforts Committee, which was housed at the Newfoundland Trade Commissioner's Office in London, Rose Marie Hodgsen described the importance of that gathering place for them. Convening there, they could collect comforts, tobacco, and news from home. She also commented on the tight-knit group who visited one another while on leave, including those with relatives overseas because their parents had met during the First World War: "They have gone visiting relations and friends. The rest of them visit each other. For everyone knows everyone else in Newfoundland. Sailors go to stay with lumbermen in Scotland. Soldiers visit the Air Force boys" (Hodgsen, 1942, p. 1). She quoted Margot Davies, secretary of the War Comforts Committee, who reassured those in the home country: "All the boys are looking wonderful. . . . They share their money and share their fun. They hang together and look each other up on leave. They are fighting fit and happy" (Hodgsen, 1942, p. 1).

"A Poetic Streak in Their Makeup"[67]: Documenting Experience and Sharing Culture

For the men of the NOFU, a large part of sharing their fun was sharing their culture—with one another at work, with visitors and one another during leisure time at camp, and with the people of the UK at whatever occasions were available. Their songs, recitations, jokes, and stories formed an extensive repertoire for self-made entertainment that alleviated boredom, drudgery, and loneliness, and presented to their hosts an important cultural dimension of the men of the unit.

In a visit to a NOFU base camp in Northumberland, a newspaper correspondent described a group of fifty-eight men at work nearby who

> sing as they swing their axes and rock like oarsmen as the crosscut saw bites deep into the vitals of old Scottish fir or Norway spruce.

67 See Ley Smith (1941).

> Above the crackle of rending timber . . . you may hear, floating down the valleys, snatches of the songs of Terra Nova—quaint melodies which tell of pick poles and peavies, of log driving and squid fishing. ("The Loggers of the North," 1940, p. 6)

The two songs referenced by the correspondent—"The Badger Drive" and "Squid Jiggin' Ground"—were iconic titles in a burgeoning folksong tradition in the home country. During the visit, Temple Butler presented the journalist with a book of Newfoundland folksongs.[68] It was a symbolic moment, indicating a cultural exchange that was woven into wartime work. The correspondent observed, as others had, that the NOFU men liked dancing, adding that "some of them give step dancing exhibitions in the recreation hut" ("The Loggers of the North," 1940, p. 6), which locals attended. Dancing, like singing and reciting, was one way the men made their own entertainment. As another correspondent visiting the camps observed of a similar evening of bunkhouse song and dance, "these young giants, partnering each other in the lancers and quadrilles, whizzed one another round, while a hundred admirers clapped" ("Lumbermen Love," 1939, p. 8).

An oral tradition of songs and stories, as well as traditional dancing, was firmly entrenched in Newfoundland at the time. During the first half of the twentieth century, folklorists from the UK, the USA, and Canada researched its rich traditions and recorded songs and stories they heard in all parts of Newfoundland and Labrador, including those that were locally composed. The early twentieth-century woods camps, in which most of the NOFU men had worked, were key places for the exchange of such stories and songs and were thus essential to their preservation (Ashton, 1986). Evening gatherings around the bunkhouse stove and weekend bunkhouse concerts were mainstays of entertainment in the isolated camps—similar to those they built in Scotland, in which men spent months at a time away from home. These practices were transplanted to the camps in Scotland and England, and the songs

68 The book was likely an edition of a Gerald S. Doyle songbook, which Doyle published in five editions from 1927 to 1978. "Squid Jiggin' Ground" was published in the second edition of the songbook (1940).

and stories, shared at concerts and other venues in the UK, caught the attention of both local people and the press.

The song repertoire included standards such as "We'll Rant and We'll Roar" (also known as "The Ryans and the Pittmans"), "The Star of Logy Bay," and "I'se the B'y," as well as fiddle and accordion tunes. Among those drawn from the unique woods-song tradition of Newfoundland and Labrador was "Twin Lakes," an early song about "subbing" or doing logging work with sub-contractors who sold wood to the large pulp and paper companies.[69] "The Badger Drive," composed by John V. Devine of Kings Cove in 1912, describes river-driving logs from the woods to the ANDCO mill at Grand Falls. One of the most popular songs of the local woods-song tradition, it held anthem-like status for the men. One newspaper correspondent described hearing it in a Northumberland camp: "From a distant hill-side came the strains of Bill Dorothy's song.[70] It tells of pick poles and peavies, and of how Bill Dorothy went down the river. It typifies the dauntless spirit of the loggers" ("The Loggers of the North," 1940, p. 6).

These and other songs were shared widely at concerts. One such concert featured a telling, two-part finale that featured two songs: "The Squid Jiggin' Ground," with men dressed in fishing garb lining one side of the stage, then "Twin Lakes," with loggers joining them on the other side. The staging was a symbolic representation not only of home industries, but of the occupational histories of most of the men, who often worked as fishers from spring to fall and loggers from fall to spring. The concert ended with "We'll Rant and We'll Roar" and then—as all NOFU concerts did—the national anthem, "Ode to Newfoundland," and "God Save the King" (*Finale*, n.d.).

The men added to this repertoire as they documented their overseas experience. Their work offered some inspiration for songwriting as well as for

---

69 For an account of the woods song tradition of Newfoundland and Labrador, including lyrics, transcriptions, historical context, and explanatory notes, see Kelly & Forsyth (2018) and Kelly (2022).

70 Bill Dorrity, whose surname was widely misspelled as Dorothy or Dorothey, was an employee of the Anglo-Newfoundland Development Company who managed the Badger drive and is named in the moniker song. In UK coverage, "The Badger Drive" was elsewhere referred to as "The Driving Song" ("Backwoods Wedding Reception," 1941).

performing. As with the woods-song tradition at home, only a small sample remains of what was likely a robust creative wartime output during a time of change. Noteworthy is "The Foresters' Song," which describes the experiences of the third contingent of 409 men who departed Bay Bulls on January 13, 1940, aboard the CPR liner *Duchess of Richmond* and arrived at Liverpool six days later. Many of the men in the group were from communities along the Southern Shore of the Avalon Peninsula, the song composers—James Carew of Cape Broyle and Patrick Carew of nearby Admiral's Cove—among them.[71] Another worker, Harry Carter (NOFU #2158) of Greenspond, wrote a handful of short unrecorded songs about the NOFU; they included "Somewhere in Scotland," which was about young men who enjoyed war work and life in Scotland yet longed for home (H. Carter, 1943). These songs, like other local compositions, were often set to popular melodies from gospel or country traditions.

Carter's compositions also included several poems with rhymes such as these from "To the Loggers Over There" that indicate just how clearly was the connection between the NOFU work and the outcome of war:

> Our Allies they are giving
> Old Hitler quite a squeeze,
> And loggers will help finish him
> By cutting down the trees.
> (H. Carter, 1943, p. 37)

Poems were often the prequel to a song, but many stood alone, such as the long narrative poem "Convoy to Scotland" written by NOFU member Angus Temple about the contingent of 963 NOFU workers who sailed from St. John's on the MS *Chrobry* on January 13, 1940. In "My Lassie from Stirlingshire," John Thomas White (NOFU #1426) captured an aspect of life that many men of the unit experienced: the love of a Scottish woman.

---

71 A recording of this song, arranged and performed by Pamela Morgan, is available on the CD *Mentioned in Song: Song Traditions of the Loggers of Newfoundland and Labrador* (Kelly, 2014).

Other examples of cultural practices that remain from the NOFU's time in the UK include the counting composition *Ten Little Lumberjills*. With references to Advie camp, the Canadian Forestry Corps, and the Culloden Moor, as well as camps at Carrbridge, it is a rare account of the relationship between the NOFU and the Women's Timber Corps (*Ten Little Lumberjills*, n.d.). The autograph book of Steven Pike (NOFU #1484), an example of the common practice of gathering wishes, blessings, and signatures from new friends and old, has many entries presented in rhyme (Pike, n.d.).

Producing NOFU concerts resulted in a multitude of creative works, most of which are undocumented. Short skits, rhymes, monologues, and songs were written to entertain audiences as well as provided valuable funds for war and other causes.[72] One concert, held at Ballater on February 28, 1941 (with proceeds divided between the NOFU Spitfire Fund and the Welfare Fund), was produced by Welfare Officer H. G. Penfold and recorded for broadcast by the BBC to audiences in Newfoundland. The broadcast featured songs, violin tunes, a cookhouse call, and greetings from individual members. According to a newspaper correspondent, "the lumbermen took up the role of stage artists with great enthusiasm, and townspeople helped them to organise" ("Deeside Lumbermen," 1941, p. 3). The recorded concert was the first of several on topics related to the NOFU; the men also participated in the "Calling Newfoundland" radio broadcast produced by Margot Davies of the BBC's London bureau (NOFA, 1945).

Sometimes local organizations delivered concerts to the NOFU camps. For example, the Scottish Forces Entertainers, whose specially outfitted concert bus visited isolated areas where defence work was being done, performed at a NOFU camp at Toxside, as described here:

> The lumberjacks—all from Newfoundland—were at the road end to welcome the party, and willing hands carried the piano from bus to canteen, where an improvised stage was lighted only by oil lamps. This setting in the midst of pine trees soon to

72 Examples of some of these short pieces are available at the Aberdeen City and Aberdeenshire Archives (PD 19 1).

> be felled at once put the artistes on friendly terms with the audience, who either sat on hard forms or stood leaning against the walls, and enjoyed two hours of bright entertainment." ("Concert Bus," 1940, p. 3)

Later in the unit's stay, after NOFU privileges were extended in 1942, concerts in local towns and villages were a regular form of entertainment.

For the men of the NOFU, their songs and stories were meant for sharing and entertaining—one camp band with a guitarist, a violinist, two drummers, and two trumpeters, played at local dances ("Tough Men," 1941). But their own concerts were more than entertainment—they provided what amounted to an informal education, a tradition of the Newfoundland camps shared in the forests of the UK as a way to build relationships and knowledge between hosts and guests. Collaborative efforts between the men and local residents deepened this meaning. For all, this sharing of culture likely inspired confidence and fostered well-being, as it demonstrated to the NOFU men themselves (as well as to others) the creativity, wit, intelligence, and acumen they collectively possessed—a counter to other more virulent representations that circulated.

And whether listening around a campfire or in a concert hall, as one correspondent reflected, these events had another level of significance:

> It was like a little bit of Newfoundland brought to Scotland and I remembered the little bit of Poland, which I saw a few months ago in Scotland, and thought of the little bits of France and Czechoslovakia and Norway where, elsewhere in this island, you will find communities from other lands bringing their own songs and their own homeland ways of life to the heart of freedom's cause. These things keep strong the ties of brotherhood among the nations of the Empire and their Allies. ("Lumberjacks: The Newfoundlanders," 1941, p. 7)

Commonalities and Differences: Relationships, Community, Culture

War and its demands did strengthen the ties that bind and created a common front where there might otherwise have existed weaker links, thereby building, in the language of the time, a wartime "brotherhood." But although many of the differences that often frayed these ties were managed for the sake of unity of purpose, they were nonetheless visible and shaped all relationships.

Despite appearance of unfettered unity swathed in cultural symbols, stories, and songs, there is also evidence of strife within the NOFU. Sometimes ethnicity, in particular, played a role in the degree of contentment some of the men experienced. This internal strife arose from systemic discriminatory attitudes and practices that existed both at home and in the UK and that are now more widely known, acknowledged, and vigorously challenged as part of the effects of colonization and cultural assimilation.

In her study of the British Honduras Forestry Unit, Angela Forbes describes how the UK authorities often infantilized and discriminated against the men of that unit. Citing official documents, she provides examples of British Honduran workers being described as "children," or "childlike," or "childish," and of how their presence represented for authorities exaggerated threats of venereal disease and miscegenation (Forbes, 2015). Such anti-Black stereotypes were especially virulent during the wartime and postwar periods, when "British" identity was normalized as White against a backdrop of increasing immigration from Commonwealth countries, including those of the Caribbean (Hanley, 2022; Waters, 1997). The film "Tree Fellers" (Bilgrami, 2004), a tribute to the wartime work of the British Honduras Forestry Unit, includes a frank discussion of the racism its members encountered. Eric Tathum, one of the men featured, spoke of spending the money they made during the early years of the war, when it was feared that German troops would reach the UK: "What's the point of me saving my money? If Hitler come, he'll kill all of us because we coloured. It was a frightening situation" (Tathum, in Bilgrami, 2004, 13:58). Several of the men remained in the UK, and married Scots women; their biracial marriages exposed the deeply rooted racism. As Eric Tathum obvserved, "Some people didn't like it at all because we were coloured" (Belgrami, 2004, 17:59).

When the NOFU arrived in the UK, descriptions of the men featured in the press emphasized their lineage, reassuring the public that their ancestors were mainly from the Devon and Dorset counties of England, with a few from Ireland, France, and Scotland, and that their dialect was "a curious sort of Elizabethan English" (Pollard, 1940, p. 4). Such descriptions positioned the men for affiliation with (and greater acceptance in) UK communities, although they nevertheless confronted both positive and negative stereotypes. The NOFU members also included men from Indigenous communities on the Island of Newfoundland, as well as from Labrador. Although their identities were subsumed within the organization, they nevertheless experienced prejudice and discrimination from the camp communities themselves.

The few snippets of memoir available provide evidence that the Mi'kmaw members of the unit—and their numbers were considerable—experienced discrimination and racism that made their time in the UK more challenging. Such overseas experiences aligned with those of working in the woods industries at home (Hanrahan, 2013; Paul, 2000). Michael Collier (NOFU #3498) of Conne River was among the last group to travel overseas with the NOFU. He worked in Scotland for three years. A former game warden and timber cruiser in New Brunswick and Labrador, he refused to discuss his time with the NOFU as part of an oral history interview decades later (Kieley, 1985). Norman Gabriel of Stephenville, who was with the first group of NOFU workers, spoke openly about how being a Mi'kmaw man resulted in alienation and cruelty throughout his life (Mishaud, 1998).

In a chapter of his memoir, *Moccasin Tracks*, Miawpukek First Nation Elder John Nick Jeddore provided an account of life as a Mi'kmaw NOFU worker in Scotland. Jeddore was among the NOFU workers who travelled to Scotland in the summer of 1941, along with other men from Conne River. He recalled how, on the voyage, they were taunted about being "Indian."[73] Throughout his four years of successful service—in various jobs that included

73 An anachronistic and inaccurate but widespread collective name for Indigenous Peoples of North America.

checker, scaler, tractor and tram operator, cook, and Lance Corporal in the Home Guard—Jeddore made many friends, yet his Mi'kmaw identity was often a point of contention for some in the unit (Jeddore, 2001).

Those who joined the NOFU from Indigenous communities on the Island and in Labrador faced new challenges at home after the war. During negotiations to craft the Terms of Union between Newfoundland and Canada in the late 1940s, the country's Indigenous Peoples were not acknowledged and, without status as Indigenous Peoples, they were not covered by Canada's Indian Act.[74] Jeddore himself was acknowledged as an instrumental knowledge holder and played a role in the struggle by the Mi'kmaq of Conne River for cultural recognition and reclamation during the second half of the twentieth century.[75] He was one of many ex-NOFU men who contributed significantly to their communities after the war.

Reports of the experiences of linguistic and cultural minorities in the unit are rare. Many of the men who enlisted, especially from the west coast of the Island, were francophone—often also of Mi'kmaw ancestry. Some francophones were mentioned in UK newspaper accounts, including Billy O'Quinn (NOFU #2533) of Millville, St. George's, whose fiddling brought him repute while he was working in Scotland (Passingham, 1941). The extent to which language shaped relationships and experiences overseas—in proximity to English and without first-language home support—and impacted first-language use and loss can only be speculated.[76] As Martin Gallant (NOFU #448) of Stephenville reflected: "I grew up speaking French. I went to an English-speaking school here, but we were all French and that is what

---

74 See Mackenzie (2010) for a discussion of these negotiations as they related to the status of Indigenous Peoples, citizenship, and the Indian Act.

75 In 2012, Jeddore was designated as a Tradition Bearer by Heritage Newfoundland and Labrador, a non-profit corporation of the Government of Newfoundland and Labrador.

76 The arrival of Americans at the US Air Force base at Stephenville accelerated language assimilation in that area. After the war, however, as a province of Canada, French language knowledge and use increased exponentially, especially following the passing of the Official Languages Act (1969), reaching close to 5 percent by the 2000s (La Fédération des Communautés Francophones et Acadienne du Canada, 2009).

we spoke" ("Back Home," 1984, p. 48).[77] The experiences of francophone men who were grouped together in camps—where they no doubt spoke only French to one another—are not well-documented. It is likely that the negative stereotypes that were used to inflict language- and culture-based discrimination at home and in the woods camps on the Island and in Labrador also followed the francophone members of the NOFU into their overseas work.[78]

International relationships and marriages also added layers of complexity. One of the most widely known elements of the NOFU's time overseas is that of the marriages between the men of the unit and the women of the UK, most especially of Scotland. The phenomenon extended far beyond the unit. Thousands of international marriages—and, later, emigrations—occurred as a result of the wartime influx of large numbers of male foreign nationals into the UK.

In early 1945, Registrar-General for Scotland J. G. Kyd reported that, since the beginning of the war and up to and including 1944, there had been 8,498 marriages between Scottish women and foreign nationals ("Canadian Beaus," 1945). Of these, 3,700 marriages with Canadians accounted for the largest number. There were 372 marriages between Newfoundland men and Scottish women. Marriages between NOFU members and Scottish women, largely from the north of Scotland, accounted for most of the marriages involving Newfoundland nationals ("Canadian Beaus," 1945). As one of the men had described themselves to a journalist early in the war, they were, indeed, "matey" ("The Loggers of the North," 1940).

---

77 Other missing stories include those of queer men, men from cultural groups such as Assyrians (whose presence was well-established in Newfoundland by the first half of the twentieth century), and other linguistic, cultural, and religious minorities present in Newfoundland and Labrador in that era.

78 The use of the pejorative descriptor "jack-a-tar" (or "jackotar," "jackietar," and other variations) to describe those of both French and Mi'kmaw ancestry was common in Newfoundland at the time (Dupuis, 2021).

Marriages Between Scottish Women and Foreign Nationals (1939–1944)[79]

Marriages	Spouse Nationality
3,700	Canada
1,586	Poland
704	Norway
581	United States of America
372	Newfoundland
335	Holland
253	Australia
170	New Zealand
154	France
Subtotal 7,855	
643	Other (nationality not recorded)
Total 8,498	

As the experiences of the British Honduras Forestry Unit reveal, both courtships and marriages were often challenged on the basis of ethnicity. For Mi'kmaw men of the NOFU, there is evidence that ethnicity was used to threaten and disrupt some relationships (Jeddore, 2001). Prevailing attitudes about various occupations, cultures, and nationalities—as well as religion and age—caused issues for various relationships. The poor reputation of the NOFU men early in the war was often a point of concern for families and communities whose women courted members of the unit, motivating a paternalism often framed by stereotypical and discriminatory attitudes.

Religious difference—most Scottish women were members of the Church of Scotland—also contributed to relationship tensions, some of which were resolved magnanimously. Grace Todd of Aboyne, Aberdeenshire, married Walter Ricketts (NOFU #1774) of Stack Cove, Bonavista Bay. Ricketts was a Roman Catholic. Todd recounted how her father, a Church of

79 See "Canadian Beaus" (1945). These statistics do not include marriages that occurred from the start of 1945 to the end of the war or, in the case of the Newfoundland number, to the end of the NOFU stay in Scotland—the summer of 1946.

Scotland elder, rather than opposing the union, advised her to convert for the sake of unity in the marriage, which she did (Stapleton, 1981). Grace Todd Ricketts immigrated to Newfoundland with her husband in the summer of 1945, and the couple and their family lived in Gander. Among her contributions to life and community in her new home were award-winning stage performances with the Avion Players.

There were times when the celebration of a marriage itself highlighted both differences and commonalities in novel, joyful ways. For example, the marriage of Stephen Walsh (NOFU #2443) of Lord's Cove, Burin Peninsula, and Catherine McIntosh of Ballater, Aberdeenshire, was celebrated at Glenmuick camp in late August 1941. Walsh was among the second contingent of recruits who travelled to Scotland in the summer of 1940. While many marriages were announced in local papers, it was the backwoods reception for these newlyweds that brought together so many facets of camp life—with a twist—that made local newspaper headlines. Tables in the camp cookhouse were adorned in white cloths for the occasion. For the dance, Foreman Kenny Gaulton (NOFU #132) sang songs performed with guitar and melodeon—with a chorus of logger guests—and fiddler Billy O'Quinn created musical harmony, playing both Newfoundland, French, and Scottish dance tunes ("Backwoods Wedding Reception," 1941).[80]

Throughout the war, having seen women employed in jobs traditionally occupied by men, members of the NOFU grew accustomed to working with shared purpose alongside the Women's Timber Corps in the north of Scotland. Not surprisingly, some working relationships grew deeper and resulted in marriages. Agnes Orman Leckie of the Women's Timber Corps, for example, married Charles Lander (NOFU #1738); upon their return to his homeland, they lived in Grand Falls (now Grand Falls-Windsor), the ANDCO town in central Newfoundland that was at the heart of the pulp and paper industry. Marion Buchan Kidd of the Women's Timber Corps married Wallace Piercey (NOFU #3211); they lived in Winterton, Trinity Bay, after the war. In a public art

---

80 Both Gaulton and O'Quinn were featured in a story about the NOFU published in *Illustrated*. A photo of Gaulton was included on the magazine cover. See Passingham (1941).

display at Botwood, such marriages between members of the two units are remembered by the image of a woman wearing the uniform of the Women's Timber Corps. "They Follow Their Hearts," a mural by Janice Udell, was commissioned by the Botwood Mural Arts Society and designed to recognize the contributions of the war brides to their new home communities.[81]

Many wives of NOFU men also became known at camp for their work service. Their roles were much less restrictive than those of the very small number of women who worked as clerks or cooks in Newfoundland camps. With the changes in welfare amenities following the unit crisis of 1942, the Dominions Office recommended hiring members' wives to operate camp canteens (Shakespeare, 1942); many also tutored the men in literacy and numeracy. Women also worked as scalers, clerks, and telephone operators. Jean MacPherson of Glasgow, wife of Larry Gladney, worked as a scaler until she was reassigned to be a telephone operator at Carrbridge headquarters (Gladney, 1999). A general shortage of labour in the UK, the advancement of women's work in various sectors during wartime, and the inclination of NOFU leadership to support their men—so many of whom started families in the UK—resulted in wives participating in numerous ways in camp life.

Although war forged such new relationships, it also, of course, separated or even divided families. Many of the men who joined the NOFU left wives and families at home. Sometimes, distance and context were used to rationalize suspending established mores, including those of marriage. As during the Great War, instances of bigamy also increased in the Second World War, with the British United Press reporting it as the number one war crime ("Bigamy," 1942). In one instance, a NOFU member was convicted and sentenced to a four-month prison term for the crime ("The *Daily News* Diary," 1942). In his own defence, the accused pled ignorance of the meaning of bigamy ("Not Heard," 1942), but *ignorantia juris non excusat*.[82]

81 The Botwood Mural Arts Society is dedicated to preserving the history of Botwood, NL, once a thriving town in the pulp and lumber industries, through the placement of a series of large-scale murals on buildings throughout the town.

82 In English, this Latin phrase would be "ignorance of the law excuses not."

Postwar Timber Supply Transition

War was a tumultuous time, challenging traditional mores, identities, and loyalties, including those of colony and nation. The end of conflict signalled the start of a new era, one that involved confronting the changes that had occurred in the war years. For the NOFU, it was a transition time, one that took them from wartime harvesting to postwar timber supply stabilization and restoration work, as well as reforestation for sustainability.

The Second World War ended in continental Europe on May 8, 1945, following the surrender of Germany signed on May 7 at Reims, France.[83] The word spread quickly, including into the NOFU camps of the Scottish Highlands. Larry Gladney recalled the night of May 7, 1945, when he and his wife, Jean, were walking by Officer-in-Charge Turner's home—Turner was in the garden preparing to raise the flag. "It's all over," he said. "We'll soon be on our way home" (Gladney, 1999, p. 69).

The end of the conflict had been in sight in the months leading to the armistice. Only two overseas forestry units remained in the UK by this time: the Canadian Forestry Corps and the NOFU. The Australia, British Honduras, and New Zealand units all left the UK in 1943. The last Canadian Forestry Corps camp closed at Beauly in mid-June 1945. As early as the fall of 1944, the Home Grown Timber Production Department had indicated they would need the NOFU to remain in the country at least six months after the end of war (Acton, 1944). Their actual stay extended into the summer of 1946. NOFU contracts officially ended when the war ended, but new contracts were issued for men who agreed to do postwar work. Approximately 1,200 men remained at first; by January 1946, there were approximately a thousand Newfoundland men in the UK. During this period, they continued to produce timber until regular supply chains were restored and a surplus timber supply from

---

83 The Act of Military Surrender was signed again the following day at Berlin, which ended fighting between the Soviet Union and Germany. The Second World War officially ended with the end of the war in Asia, which occurred on September 2, 1945, when Japan signed the Instrument of Surrender at Tokyo following the atomic bombing of Hiroshima and Nagasaki.

Germany could be established (Lloyd, 1945). Some also participated in cleanup and reforestation work, replanting the forests they had felled during the war (T. V. Curran, 1987).

An eagerness to see the end of war and return home sometimes resulted in disciplinary issues. Just a little over a month before war's end, Turner issued a memo to the unit, reminding those who might be "getting the idea that the war is over and that people can slack off" that there was still work to be done and "the demand for timber was greater than ever before" (Turner, 1945a, p. 1). The notice followed the imprisonment of some NOFU men for breach of peace and property damage, and the firing and repatriation of three others who refused to work. Turner admonished what he called these "weak sisters" (Turner, 1945a, p. 1) in the unit whose behaviour tarnished its reputation and that of the country of Newfoundland—a reminder that harkened back to the unit's initial years in the UK. But Turner also gestured to the future and the postwar road ahead in Newfoundland, offering the men a challenge: "We don't want any slacking on the job, or hell-raising off it. . . . Newfoundland needs every bit of good advertising it can get [so] let's provide that advertising by proving beyond doubt that we are the best logging outfit in the United Kingdom" (Turner, 1945a, p. 1).

By early March 1946, plans to repatriate the unit were in place. As the month ended, no new contracts were issued and members remaining in the UK were released. Some remained with the unit until transportation became available. By the end of July 1946, the unit was fully repatriated ("Forestry Unit Will Be Disbanded," 1946). The last group of NOFU members, including Jack Turner, departed Southampton aboard the SS *Lady Rodney* on July 19, arriving in St. John's on July 26. Aboard were 153 NOFU members plus 19 wives, 3 children, and 47 Royal Navy and 4 Rescue Tug Service members ("SS *Lady Rodney* Due," 1946).

On July 22, 1946, while the last group of repatriated workers were travelling the high seas between Southampton and St. John's, Viscount Addison, Secretary of State for Dominion Affairs, expressed his appreciation for the work of the NOFU, an excerpt of which was published in *The Evening Telegram*. Addison wrote of "the invaluable service of these men through the war years and in this first year of peace" and "the real contribution to victory which they have made by their cheerful and unfailing assistance to us in our desperate need for timber during these years" ("Expresses Thanks," 1946, p. 3).[84] The NOFU had been the first unit on the ground when war began and they were the last to leave, closing a chapter in the unit's wartime service but by no means ending its story.

During the period of service in the UK, a total of thirty-four NOFU members died of various causes, including work-related accidents. In each case, co-workers funded a gravestone. In a letter to Alfred Hart, Emily Butler of Port Rexton, who lost her son James Butler, likely reflected the feelings of other families with a relative lost while working with the NOFU:

> I am glad to know that he has a stone to mark his grave . . . It certainly consoles me and I shall always feel so grateful to them. Although I cannot visit the spot where he is laid, yet my thoughts wander there quite often. (E. Butler, 1941, p. 2)

In 1945, the NOFU requested that Trade Commissioner for Newfoundland D. J. Davies ask the Imperial War Graves Commission to maintain the thirty-three graves of the NOFU men, as they did those of military service personnel.[85] Davies subsequently requested permission from Home Affairs in Newfoundland to approach the Commission and, in case they refused the request, propose they tend the graves at the expense of the NCG (Davies, 1945). The Imperial War Graves Commission agreed to the latter proposal.

The graves of the NOFU men who died in the UK are tended today by the Imperial War Graves Commission. The NCG bore the cost of upkeep until 1949, when Newfoundland joined the Confederation of Canada. Neither

84 Viscount Addison was Christopher Addison.

85 Two NOFU members who died in England are interred in the same grave at Carlisle. For an account of these thirty-four deaths, see Green (2025).

NOFU benefits nor the care of members' overseas graves was secured during the negotiation of the Terms of Union between Newfoundland and Canada in 1948. The attitude of indifference for the war graves of the NOFU men who died overseas was similar to general attitudes that followed the NOFU after the war. The status of their war service was nebulous; consistent with its experiences while overseas, it would be up to the men themselves to make the case for proper recognition and concessions. In 1953, following discussions between the Province of Newfoundland and the Government of Canada, the federal government agreed to cover costs previously paid by the NCG and then the country's newest province.

SECTION THREE

REPATRIATION AND REINTEGRATION: POSTWAR STRUGGLES

As the war continued in Europe and Asia, the economy of Newfoundland was being transformed by related events. Both the United States and Canada had developed a strong interest in the country during the war because of its location: as North America's most easterly point, Newfoundland was a gateway to the continent. Each country wanted to ensure it was militarily protected at this key entry point. Agreements for the construction of American and Canadian armed forces bases to optimize defence provided thousands of jobs on the Island and in Labrador. At the height of the construction boom in 1942, almost twenty thousand residents were employed in some aspect of base construction (High, 2003). Demand increased in existing and new spinoff industries, as well. Wages went up and the general standard of living vastly improved.[1]

As Newfoundlanders and Labradorians observed American and Canadian soldiers in their communities, they began to want and expect more for themselves, their families, and their country. Internal migration from rural communities to towns where the bases were located, as well as out-migration to Canada where the economy was expanding and labour demand was high, also increased. In the pulp and paper industry itself, labour became increasingly scarce—woods workers had other employment options and at higher pay (High, 2003; Hiller & O'Brien, 2019; Neary, 1988).

The economic boom was so dramatic that the NCG, which required annual grants-in-aid from the UK government to maintain its operations, began

1 See High (2003) for a discussion of the economic impact of these bases.

Woods workers with bucksaws, axes, and shovels at the Canadian military base construction site, Goose Bay, Labrador, 1941. (Courtesy of Them Days Archive, Jane Lethbridge Collection, M181)

to run large surpluses that allowed it to provide interest-free loans of a total of over $10 million to support the UK war effort (Noel, 1971). Even though the prospect loomed of another postwar recession, which somewhat tempered confidence, this increased economic prosperity intensified discussions about the end of government by commission and the political future of Newfoundland that would follow it.

Countering the infusion of optimism created throughout the war was the uncertainty about which direction the country should go when the opportunity to decide its future arose. The NCG was unpopular, yet a return to responsible government created anxiety for many Newfoundlanders and Labradorians. At war's end, Confederation with Canada was not yet considered an option—there was virtually no support for it among the population (Hiller, 1997a).

It was to this transformed economic and social context and uncertain political future that those who went overseas in various capacities during the war, including the men of the NOFU—hundreds of them with wives and children—returned.

Immigrations

During the wartime period of prosperity and with the presence of American and Canadian personnel, Newfoundlanders and Labradorians became more familiar and comfortable with foreign nationals living among them. Lifelong friendships were formed, many marriages took place, and when the war was over many women and children emigrated to new lands with their foreign husbands. These changes at home were in step with those occurring overseas, where Newfoundlanders and Labradorians serving in various capacities were getting ready to return home with wives and families after the war. These relationships and marriages cemented bonds that bridged differences and reshaped identities.

After the war, most of the Scottish women who married NOFU workers immigrated with their husbands. They were part of a significant wave of approximately eight hundred "war brides," mostly from the UK but some from other European countries as well, who married wartime personnel from Newfoundland and eventually made the country their home (Casey & Hanrahan, 1994). While there is no official list of their names, Jackie Sheppard Alcock has compiled research identifying approximately six hundred such women (Sheppard Alcock, 2015).

The available stories of the women who immigrated and stayed in Newfoundland or Labrador—by some estimates as many as 25 percent did not remain—share common themes that are focussed on the differences between the new country and the old country (Casey & Hanrahan, 1994). The war brides dealt with numerous challenges related to the absence of amenities they had taken for granted in their previous lives—electricity and indoor plumbing among them—as well as rudimentary means of transportation and limited access to consumer goods. There were also cultural differences that ranged from house building and meal preparation to social celebrations and language usage (Barrett et al., 1996).

Documented snippets of their experiences often reveal a cheerful optimism about these differences. In 1946, Myra Cribb of Aviemore, Scotland, who married Thomas Cribb (NOFU #1614) of Head of Bay D'Espoir, wrote to *The Sunday Post* in Dundee, Scotland, from her new home. She told readers about the bright wooden houses designed and built by the men themselves that were "much warmer that our concrete ones" and the modes of transportation: longer trips by steamer or motorboat, and shorter jaunts by dory in summer and dog and sleigh in winter—the "only things on wheels being prams and horse carts" (Cribb, 1946, p. 4). Jessie Blackmore of Whitburn, Scotland, who married Harland Blackmore (NOFU #2364) of Pinchard's Island, Bonavista Bay, wrote similarly to the *West Lothian Courier* in Bathgate, Scotland, noting that although the people of her new home were "behind the times with lots of things," they had made her feel welcome and she felt it unlikely she could meet "a nicer lot of people" (Blackmore, 1946, p. 2).

There were other shared themes of perseverance and resilience inspired by love and strengthened by family and community. The women drew on the natural beauty and personal warmth that are great endowments of their new home, and they often let humour be an antidote for difficulty (Barrett et al., 1996). For many of the war brides, differences between the new and old countries also became opportunities for service. Rather than bemoan what was lacking, they set about to provide it. Margaret Crocker, who had married Walter Crocker (NOFU #3506) of Heart's Delight, commented on another consistent theme of the war bride narratives—the decision to marry and immigrate: "Would I do it all over again? As I sit and think about it, the love and the laughter, the tears and the happiness, yes I would" (Crocker, 1996, p. 23).

Based on almost two dozen interviews, George Casey and Maura Hanrahan (1994) concluded that many of the war brides of Newfoundland and Labrador had led extraordinary lives, both before and during their marriages, in their home country and in their new home. Some of these women received great accolades for their contributions—Barbara (Micklethwaite) Barrett, Christina (MacLellan) Cole, and Morag (MacKinnon) O'Brien, for example, received the Order of Canada. Most led less celebrated lives but nevertheless transformed, in smaller and greater ways, the communities of which they were a part by sharing with locals "new ideas and outlooks" (Emerson, 1942, p. 4), while also adapting to other ways of living found in their new locales.

Lady Eileen Mary Walwyn, wife of Governor Humphrey Walwyn (front row, fifth from left), with some of the war brides and their children at Government House, the official residence of Newfoundland's governor, St. John's, 1946. (Courtesy of The Rooms Provincial Archives Division, VA 15b–55.2)

Many of the men of the NOFU who married while in the UK chose to remain there after the war. Edward J. Flynn (NOFU #223) expressed some of the reasons for this choice, which included the preference of many of their wives to remain at home, as well as the availability of educational opportunities for their children and the prospects for the men themselves when measured against the unknowns of Newfoundland's future. Regardless of the choice, Flynn encouraged mutual understanding among the men, emphasizing that, "wherever we may decide to settle down we shall do so with confidence and a determination to do our best for ourselves and our families" (Flynn, 1946, p. 2).[2]

2 Like the war brides who crossed the ocean with new husbands, many of the men who remained in the UK contributed, often significantly, to their new country. For example, Colin Ploughman (NOFU #2340) of George's Brook, Trinity Bay, was appointed to the Most Excellent Order of the British Empire in 2009 in recognition of his conservation work at the Big Burn Walk, Goldspie, Sutherland. Lawrence Ryan (NOFU #2934) of Port Saunders became a multi-millionaire in Wales by salvaging and selling coal tailings (slag) as commercial fuel.

Most of the NOFU men returned to their homes in Newfoundland and Labrador—and many felt the trepidation and some sense of displacement that migration creates. One NOFU worker questioned what awaited them thus:

> What are we to do? Where do we fit in? Six years away from a country and a man returns as a stranger. All his friends are gone or changed, his fishing site, his farm or his business has become decayed and will take years to become normal again, if ever, and his job is filled by another person who is up to scratch on the business which you have forgotten during these years. (J. G., 1945, p. 2)

Inspired and emboldened by their experiences in the UK, others expressed a determination to seek change. One NOFU member speculated on the conditions under which "the seed of democracy [can] germinate" in Newfoundland and grow politically minded people (Laracy, 1945, p. 2). Another wrote of the poor conditions in northern Newfoundland, the area to where he would return, confessing that "after living five years in [Scotland], and seeing conditions here, I, myself, sometimes think it is incredible" (Traverse, 1945, p. 2). Once at home, such thinking likely aligned with interest in Confederation with Canada, once it became an option and given the prosperity it promised on so many fronts.

Another group of men initially stayed in Scotland for an extended period after the war and then returned to a transformed Newfoundland. Martin Gallant, a francophone from Stephenville, where a US air base had been built during the war, spent a year with the unit before transferring to the Royal Air Force. His brother William (NOFU #447), who had joined at the same time and also transferred to the armed forces, had been killed during the war, and three of his cousins died on the HMS *Hood*. Gallant remained in Scotland after the war. He learned carpentry and with his wife raised a son and two daughters. On a home visit after nineteen years away, and with his old country now part of Canada, he and his family decided to stay. His Scots wife, Mary, observed of Newfoundland that "not too many people who come here and stay for any length of time want to leave" ("Back Home," 1984, p. 48). Gallant reflected on the differences he had observed between the pre-war country of Newfoundland and Newfoundland as a province in the Canadian Confederation, in particular, around interdependence: "People are better off, but something is missing. In the old days if you were in trouble everybody helped. . . . People depended on each other" ("Back Home," 1984, p. 48). Such observations exemplify a common migrant experience: separation from the homeland, combined with reflection and change, casts it in a new light.

Other expatriates had only rare opportunities to return. John Reddigan of Stirlingshire was a child of a Scot-Newfoundlander marriage. His father Richard ("Dick") Reddigan (NOFU #2239) of Calvert, was posted to Rosemarkie on the Black Isle peninsula upon arrival in the UK; he would marry and remain in Scotland. John recalled growing up in a home in which the songs of Newfoundland filled the air (Reddigan, 2009). In 1975, Dick Reddigan returned to Newfoundland for the first time since leaving in 1940. It was also the first time his son John met his Newfoundland relatives. Many years later, he reflected on the strong bond they had formed and his pride in his Newfoundland heritage (Reddigan, 2009). The prolonged time between departure from home and a return visit was an often-poignant part of the immigrant experience that was shared both by many women who left the UK and many ex-NOFU men who remained there.

At the heart of these immigrations were relationships, many of which were built on the abiding ties of ancestry and affection. The wives who immigrated to Newfoundland were a tangible reminder of the connection between Newfoundland and the UK. Their efforts to maintain a connection among themselves based on their home of origin and the shared experience of immigration—as demonstrated in meetings, reunions, and newspaper and book publications (Barrett et al, 1996)—also sustained imperial ties in the new Canadian province while building affection for their homeland. As Ian Cameron of Ballater observed about the connections of history and culture that now exist between Scotland and Newfoundland and Labrador: "We have these links and we cherish them" (O'Brien, 2005).

Civil Re-Establishment and Postwar Newfoundland

The men of the NOFU who returned to Newfoundland, with or without families, found little official support available to them for their transition to postwar life at home. Despite their essential contribution to the Allied war effort, their very marginal place in civil re-establishment efforts became clear as the end of the war approached.

Commissioner of Home Affairs and Education Albert J. Walsh announced the NCG plan for returning service personnel in a radio address to the country broadcast on March 14, 1945, and published the following day in the St. John's newspapers. The plan was subsequently also published by the NCG in the pamphlet *When You Come Home* and later reprinted as *Now That You Are Home*. In 1943, a committee of the NCG and the Board of Trade agreed to a series of benefits for those who served overseas that ranged from deferred pay and clothing allowance to supplemented dependent and pension allowances, travel expenses, and assistance in employment. While the NCG was quick to distinguish its civil re-establishment plan from its reconstruction efforts, their common objectives were clear. The principles that guided the plan emphasized providing opportunities for increasing education and technical competence in keeping with labour market demands, encouraging returning workers to engage in primary industries in the country, and easing their transition into civilian life through supports that offset financial strain (A. J. Walsh, 1945).

Central to the plan were educational benefits that enabled service personnel to resume interrupted study at any level. The plan highlighted teacher education, with terms to encourage teachers who had enlisted to return to the classroom and new recruits to join the profession. Of those teachers who had served, Walsh commented: "Men with the experience and broad outlook which they have acquired would be a great asset to this profession and would make a profound impression upon the life of any community in which they might reside" (A. J. Walsh, 1945, p. 2). Incentives and allowances were available for education at all levels of schooling, including correspondence and evening courses. Vocational guidance by specially trained members of the

Division of Civil Re-Establishment was also made available, and a vocational training institute for veterans was established.[3]

The plan also offered assistance to men wishing to return to fishing. To build capacity in fisheries leadership and research, the plan included allowances for university education (for a small number) in marine biology and chemical engineering. Incentives for individual small businesses were included, as were unemployment, disability, and medical benefits. The plan extended to housing considerations and land settlements and covered agricultural training. These courses—some farm-based, some at the Agricultural College in Nova Scotia—were provided as preparation or upgrading to participate in land settlements in the Humber Valley. The country's second major industry—forest-based production (the woods-based industries of lumbering, logging, and pulp and paper)—received no mention in relation to study, business incentives, or even re-establishment more generally. Thus the plan reflected and continued the hands-off approach the NCG had taken during the war toward the pulp and paper companies; the industry was experiencing a postwar labour shortage, and the companies wanted nothing to compromise their access to workers.

In his speech, Walsh was not explicit about whom the plan had been designed for beyond referencing "present and former members of our Armed Forces" (A. J. Walsh, 1945, p. 2). In a radio interview that followed on March 19, 1945, J. A. Cochrane, Director of Civil Re-Establishment, addressed a series of questions related to the plan. When an interviewer asked if the plan applied to all those who served in any capacity, he replied that only those who served in the UK armed forces—navy, army, and air force—were eligible. He added:

> It is recognized that the Merchant Navy and the Overseas Forestry Unit made a great contribution, yet these men were paid at civilian rates of pay and should have been able to save sufficient money to re-establish themselves in civilian life. Many of

3 The Vocational Institute for Ex-Servicemen, located in a former Canadian Navy machine shop on the south side of St John's, opened on April 24, 1946. Alfred E. Hart, former Administrative Officer of the NOFU, served as director for four years ("Vocational Institute," 1950).

> them will continue their employment at sea or in the woods. (Cochrane, 1945, p. 5)

It was a version of the response provided directly to the NOFA's Rex Browne (NOFU #1129) by the Commissioner of Natural Resources, Peter Dunn. On behalf of the unit, Browne, who had been an Assistant Superintendent in the NOFU, wrote to Dunn to express members' strong objection to their exclusion and their concern about the deposit required of them should a place become available in the settlement plan—the only one for which they were eligible—and the availability of work upon return to the country (Browne, 1945). Dunn replied with a reassurance about the shortage of loggers and to impress upon Browne that, even with the deposit required of an ex-NOFU member who might participate in land settlement, the remaining amount granted by the NCG was still "a generous contribution to members of the unit and much more than is granted to civilians" (Dunn, 1945b, p. 3). And therein was the rub. The men of the NOFU were overseas workers—civilians. As the NCG saw it, they had been *employed* in war work, they were not armed services personnel.

Earlier in 1945, in an address to the St. John's Board of Trade about questions he had received from members of the NOFU and the armed forces, Commissioner Dunn complained how members often asked about postwar Newfoundland, sometimes expressing the feeling that if things had not improved, they would not return. "My invariable reply has been that the Newfoundland of the future will depend upon what the young men make it. If they do not return firmly resolved to remake the fortunes of their country then it can hardly be expected that somebody will wave a magic wand and do it for them" (Dunn, 1945a, p. 10). It was another jab at people whom the NCG saw, for their reputed lack of initiative, as a liability that stunted growth in the country. It is possible that the issues and concerns raised about the unit during its wartime work in the UK were also reflected in his attitude.

In the same address, Dunn spoke at length about the importance of forest industries—the pulp and paper industry in particular—and its labour needs, estimating the two large companies hoped to employ over eight thousand men in 1945, and that their plans for expansion "depend on the support they receive from labour" (1945a, p. 12). Dunn also emphasized the need to regulate cutting through local management, to practise reforestation, and to educate communities about forest resource protection—all of which were important opportunities for education, growth, and management in a postwar context. He was well-aware of the significance of these industries; the issues with forest management in relation to them had been outlined in the *Amulree Report* (NRC *Report*, 1933). Despite the newly acquired skills the ex-NOFU men could bring to these forest priorities, the NCG had no official plan for re-establishing the wartime overseas woods workers beyond an informal alignment with the pulp and paper companies regarding reintegrating the men into their existing labour needs.

Along with the possibilities of employment at the two large Island-based companies—the ANDCO and Bowater—there was work in Labrador, including at the contentious Labrador Development Company at Port Hope Simpson. Established at Alexis Bay in 1934 by John Osborne Williams, a friend of then Commissioner of Natural Resources John Hope Simpson, the company had been plagued by financial and labour issues that eventually led to a 1945 inquiry by Justice Brian Dunfield. Given an opportunity by the NCG to redeem itself and salvage the pit prop operation following the inquiry, Williams attempted to recruit workers from the Island, the returning ex-NOFU workers among them ("Wants Loggers," 1945). Despite his efforts, the company was liquidated in 1947. John Grieve also recruited ex-NOFU men for the Three Rapids Estate Lumber Company, his family operation at Kaipokok Bay, Labrador, but it also folded shortly after the war (Munro, 2011a, 2011b). Despite the many ex-NOFU men who went to Labrador, among them leaders such as camp foreman Kenny Gaulton of Grand Falls who worked at Three Rapids Estate, Labrador woods industries did not provide large-scale or sustainable postwar employment.

The Postwar Land Development Policy

Those who drive through present-day Cormack, Newfoundland, a small community a short distance north of Deer Lake on the Great Northern Peninsula, may be unaware of its origins in the postwar period or of its original name: the Upper Humber Land Settlement.[4] This area was one of the locations chosen to settle veterans into farm allotments after the war with the goal of increasing agricultural production in the country.

The postwar land development plan was not the NCG's first settlement plan. It had created farm initiatives in the 1930s to ease the population's dependency on a failing fishery and provide increased food security for families.[5] As in the development of the logging and lumbering industries, new communities arose from these initiatives on the Island: Markland, Haricot, Lourdes, Midland, Brown's Arm, Winterland, Sandringham, and Point au Mal. The postwar land development policy for civil re-establishment added one more community—Cormack—to this list.

Overseen by the Land Development Division of the Department of Natural Resources, this aspect of the NCG's civil re-establishment plan was the only one that allowed for involvement of ex-NOFU members—but not in equal standing with armed forces veterans. The plan was initially two-tiered: Plan A offered farm allotments of up to 350 acres in the Upper Humber area and Plan B offered spaces for farming in or near the home communities of service personnel. Plan C, introduced later, placed men on vacant holdings in the initial 1930s land settlements.

Plan A's offerings were for veterans of the armed forces. Ex-NOFU were eligible to apply for an allotment only if veterans did not fill all of them. Its benefits included a ten-acre cleared portion of land, a six-room bungalow, $600 for barn construction materials, and $724 for the purchase of livestock, farm tools, seeds, and fertilizer, in addition to training and transportation bonuses. Farm training was completed in Nova Scotia, New Jersey, or at the Demonstration Farm at Mount Pearl ("Annual Report Land," 1947). By August 1949, ninety-six farms had been established under this plan.

The more modest Plan B was for start-up farms on arable land that an applicant already owned or intended to acquire. It provided for clearing an initial ten-acre area and included a $724 allowance for livestock, tools, seed,

4 In 1947, the Farmers Association in the area voted to rename the community Cormack in recognition of William Eppes (Epps) Cormack.

5 For an overview of these initiatives and the resulting issues, see Higgins (2007a).

and fertilizer. Farms developed under this plan were in all parts of the country. Ex-NOFU members could apply for funding under Plan B if they met certain criteria: they had remained with the NOFU until VE Day or later, or they had been repatriated due to illness or for compassionate reasons. If successful, they were required to contribute $250 before benefits would be given ("Annual Report Land," 1947). By the end of 1951, 184 families had been supported under this plan, of which thirteen were ex-NOFU members ("Land Clearing," 1951).

Under Plan C, applicants were assigned existing allotments in the older land settlements and given $724 allowance for tools, seed, fertilizers, and other essentials ("Annual Report Land," 1947). Seven families were settled under this plan.

The terms of the postwar land development policy were not well-received by the NOFU, who were watching them closely from afar. As at the start of the war and throughout, by war's end the NCG still showed no inclination to invest in the unit—to plan for the future of its members or acknowledge their contribution to the war. The NOFA position, as indicated in the editorial of the July 1945 issue of the *Newfoundland Foresters' Bulletin*, was candid: "This attitude of the Government is deplored by members of the Forestry Unit" ("Editorial," 1945b, p. 1).[6]

A Forgotten Service: Reintegration Without Benefits

Confronting the attitude of the NCG and striving to obtain benefits became the postwar focus of many ex-NOFU members. Not only had their service been ignored and invalidated; many on the Island and in Labrador knew little about the unit, its experiences, or its accomplishments. Unlike the Newfoundland Forestry Companies of the First World War, whose scribe, Quartermaster Sergeant John A. Barrett, had penned regular accounts of the unit's work in Scotland for publication in Newfoundland newspapers (Kelly, 2020), the NOFU

6 See also "Foresters Not Included," 1945

had received no concerted coverage, no doubt, during the early years at least, because of its beleaguered status. Short news items, a small handful of letters from individual members, reprints of some items published in UK newspapers and the *Bulletin*, and coverage of official visits to the unit by Newfoundland and UK authorities, along with two published updates from Officer-in-Charge Jack Turner (in 1943 and 1945) largely account for the total coverage of the NOFU in local Newfoundland newspapers during the unit's six-year run overseas.

Things had been somewhat different after the First World War. The Newfoundland Forestry Companies had been a uniformed unit of the Newfoundland Regiment and thus were eligible for allowances and pensions along with their compatriot soldiers. Still, they found themselves fighting for parity with the Regiment in some areas, sparking debate in local newspapers and revealing deeply embedded attitudes about hierarchies of service. Even within the Regiment there were questions about their right to any medal, despite demonstrated eligibility.[7] But from the perspective of the NOFU after the Second World War—who were ineligible for any pensions or allowances—the Newfoundland Forestry Companies had been recognized and rewarded for their service by the Government of Newfoundland *and* they were thus included in benefits negotiated for veterans through Confederation with Canada. The NOFA would cite them as one of the wartime service groups—along with the Canadian Forestry Corps—with whom they sought parity.

The executive of the NOFA, who led initiatives for better recognition for the men, struggled in their own transition to life and circumstances in Newfoundland. In late 1945, in an effort to reconnect with repatriated members, the association publicized an advisory notice in local newspapers announcing its head office at Glovertown ("Notice," 1945). Plans to operate at home, including resuming the publication of the *Bulletin* under the editorship of Thomas Curran, were announced in spring 1946 ("Foresters Association," 1946).[8] The NOFA executive, with delegates from around the Island, met at

---

7 See Kelly (2020) for a discussion of this issue.

8 In 1995, Thomas V. Curran was awarded an honorary Doctor of Laws from Memorial University of Newfoundland in recognition of his distinguished wartime and postwar work on behalf of the men of the NOFU. See J. Scott (1995).

St. John's on September 28, 1946—the first time since repatriation. Their plan was to open an office in the capital city and develop branches throughout the Island. Membership was at approximately seven hundred at the time ("Ex-Foresters Association," 1946), less than 20 percent of the maximum possible enrolment of almost 3,600 but almost 60 percent of the NOFU's UK strength of 1,180 at the end of 1944. There was work to be done to reconnect ex-members of the unit for what lay ahead, and to profile the work the NOFU had done overseas. As Curran later commented, by the time the unit returned home, "we were hardly known in Newfoundland" (Canada, Parliament, House of Commons. Standing Committee on Veterans Affairs, 1960, p. 103).

It would be well over a decade, however, before any branches were formed. Membership drives and branch formations did not occur until the period when NOFA executive were testifying before the House of Commons Standing Committee on Veteran's Affairs in 1960. During those hearings, it became apparent that the organization lacked both funds and, understandably, given the short period of being Canadian citizens, adequate knowledge of the workings of the Government of Canada. When asked about the delay in appearing before a Veterans Affairs (VAC) standing committee at meetings in 1952, 1954, and 1956, Curran replied that the NOFA dealt directly with the Minister and Deputy Minister of Veterans Affairs, adding, "We were not aware, our association, . . . that there was such a thing as a Veterans Affairs committee" (Canada, Parliament, House of Commons, Standing Committee on Veterans Affairs, 1960, p. 104).

It was a telling point. The NOFU men were representative of Newfoundlanders and Labradorians at large. Their experience of suspended democracy had resulted in diminished citizenship and reduced their knowledge and experience of how to represent their voices procedurally and collectively. The Liberal Member of Parliament for Burin-Burgeo, Chesley Carter, made the point years later in the House of Commons as the struggle for ex-NOFU rights continued to unfold. When the Progressive Conservative Member for St. John's East, James McGrath, asked him about NOFA delays in approaching VAC, Carter replied: "For fifteen years Newfoundlanders had no representatives at all; they had no vote, and they were not in the habit of referring problems to their members.

They were not in the habit of even thinking in those terms" (C. Carter, 1962, p. 863). Without their Officer-in-Charge to present their case, their Members of Parliament silent or uninformed about their contributions, and a Premier (Smallwood) indifferent to their needs, their case lay idle for many years.

Indifference to the plight of the ex-NOFU was built on a lack of wartime attention at home, including what Thomas Curran described as the "complete apathy" of the NCG in regard to the welfare of the men, which rendered them "the forgotten service" (T. V. Curran, 1987, p. 66). It was certainly a widespread belief, among the men, that they had been forgotten or ignored when the NCG plans for reintegration were announced. In a series of Letters to the Editor of the *Bulletin*, assembled under the title "Foresters Get Last Preference," members took exception to the hierarchy of support for wartime volunteers and the place of the NOFU workers within it. A member who worked at Revack camp called for a fair deal, arguing the NOFU should not be the last considered for support, and noting that "we were not the last to be called upon for overseas service, we were not the last to answer the call; so why should we be the last to share fair in the Government's Postwar Plans" (J. T., 1945, p. 1). One member at a Carrbridge camp felt the NOFU workers were left "collecting the crumbs" (J. G., 1945, p. 2). Temple Butler, submitting his thoughts from the Deishar camp, wrote that, as well as facing dangerous waters, bombing raids, and wartime causalities, "we have spent a longer period of exile than any other Allied group" and "have a right to recognition and status" (T. Butler, 1945, p. 2).

The inattention and apathy had real policy consequences when the NCG began its postwar planning. Positioned in the public eye as waged workers who had saved considerably while away, the NOFU men likely garnered limited public sympathy in their push for fairness. As Commissioner Dunn revealed: "The proposals for members of the Forestry Unit have been drawn up in the knowledge that they have received a much higher rate of remuneration than the men in the Armed Forces, and have been able, in many cases, to save fairly substantial amounts" (Dunn, 1945b, p. 3). Dunn failed to acknowledge that the NOFU labour rates had not kept pace with their counterparts at home, whose wages had tripled during the war. Their demonstrated commitment to savings to support the war was also turned against them.

The omission of pension and allowance clauses in the contracts signed by NOFU workers presented a huge obstacle to postwar recognition at home. Their status as civilian workers, not soldiers, exacerbated these difficulties. The unit found itself arguing, without success, for postwar recognition from a weak, after-the-fact position with a government that was arguably in lame-duck mode pending a decision on the future of Newfoundland. Its refusal to acknowledge the NOFU men beyond its initial and very limited terms created additional obstacles for recognition once Newfoundland became a province of Canada. Without carry-over rights, the path to recognition by the Government of Canada was not a clear or easy one.

NOFU workers who were injured while working in the UK and endured long-term hospitalization as a result of those injuries received no support upon their return. Roderick T. Colbourne (NOFU #2826) was hospitalized in Scotland for eighteen months with a foot injury that left him permanently disabled. Yet when he returned to Newfoundland after five years of service with the NOFU, there was no program or pension available to him (Colbourne, 1961). Extended hospital stays such as Colbourne's were not unusual—woods work was dangerous and injuries could be severe.[9] While the NOFU contract stipulated that medical care and compensation for lost wages were covered under the UK Employers' Liability Act and Workmen's Compensation Act ("Provision," 1942), these benefits ended when the men returned home. A lump-sum payment to a worker on discharge was both the capstone of the benefit and the end of the UK government's liability.[10]

NOFU members who chose repatriation faced a very different set of circumstances than those who chose to stay in the UK. Concerned to know what the benefits were for members who wished to remain overseas, Turner raised the issue with the Dominions Office in May 1945. By then, NOFU workers had been apprised of their status under the NCG Re-Establishment Plan. The Dominions Office confirmed that the Ministry of Labour and National Service considered

9 Several examples of NOFU member hospitalizations, ranging from twelve to thirty-four months, were reported ("News Budget," 1944, p. 9).

10 Several letters like Colbourne's are on file at The Rooms Provincial Archives, St. John's. See, for example, GN 1/3/A, Box 251, Files 1 to 2, Forestry.

that the work of the NOFU met "the conditions of effective service" (Rouse, 1945, p. 1). Men remaining in the UK were thus eligible for postwar educational training and other opportunities offered by the UK government, specifically under its Vocational Training and Further Training and Education programs.

Members who were under contract with the Ministry of Supply for postwar timber production and other tasks were required to wait until their contracts were terminated before enrolling in any training program. Alfred Hart raised the issue of educational eligibility of the NOFU members, given "the educational standard of the average member of the unit may not be up to the standard required by the Ministry." He was reassured that "sympathetic consideration would be given by the selection committee, particularly in the case of otherwise well-recommended men" (A. E. Hart, 1945, p. 1). The tenor and spirit of these responses were in marked contrast to those from Dunn and other members of the NCG.

For its part—and unsurprisingly—the NCG had no objection to the Dominions Office notifying the NOFU men of the postwar opportunities in the UK, as long as it was made clear that members desiring such training had to apply under the UK programs (Newfoundland Commission of Government, 1945), an emphasis that underscored they would have no such access at home.

Referenda: British Subjects or Canadian Citizens?

The uncertainty about the direction in which the governance of Newfoundland would unfold further complicated the repatriation and re-establishment of returning overseas wartime personnel. The process by which that direction was chosen had several steps and was not without its own behind-the-scenes motivations and divisive activities.

The mechanism for choosing Newfoundland's political future began, officially, on September 1, 1946, with the opening of the National Convention. Its forty-five delegates had been elected to represent thirty-eight districts, one of which was Labrador and the thirty-seven others were various areas of the Island. The men who assembled in St. John's (and they were all men) were considered representatives of the population; their task was to examine the economic changes in Newfoundland since the appointment of the NCG and make recommendations—in a non-binding, advisory capacity—to the UK government of their preferred ballot choices for a national referendum on the country's political future.[11] Ten committees were struck to deliver reports on various topics related to the economic and social future of the country; the Finance Committee used this information to present an economic outlook.

Two delegations from the National Convention were struck to explore options for the political future of Newfoundland, one to visit London and the other Ottawa. The UK government reinforced its position to the London delegation (whose members were not inclined to favour Confederation with Canada): no aid would be available should the country vote a return to responsible government and later find itself in economic difficulties. The UK coffers were compromised after a costly war and the government felt Newfoundland was likely to be a liability.[12] It was thus much more interested in divesting itself of responsibilities for Newfoundland; Confederation with Canada was the preferred vision for its future.

The delegation to Ottawa—which included pro-Confederation delegate Joseph R. Smallwood and the Convention's Chair, F. Gordon Bradley[13]—was able, on the other hand, to manoeuvre draft terms of union. With an increasing awareness of Newfoundland's strategic importance in North American security came renewed Canadian interest in the area on its eastern flank—it could be seen as a crucial part of the federation. Canada was

---

11 Noteworthy for the NOFU was the election of Alfred W. Watton (NOFU #930) of Change Islands, a teacher and owner of a small business at Fogo. Watton had transferred to the navy at the end of a six-month contract. He was elected to the National Convention by acclamation to represent Fogo. Like many of the NOFU men, Watton supported a return to responsible government (Watton, 1947).

12 An NCG development plan proposed in 1944 to the tune of $100 million had been soundly rejected by Treasury as unaffordable (Neary, 1983).

13 Despite his work on behalf of loggers in 1934 and his knowledge of their labour standing, Bradley did not emerge as a voice of support for the NOFU during negotiations of the Terms of Union.

also playing offence against any emerging moves by the United States and was willing to show a generous hand. Nevertheless, back at the National Convention, including the option of Confederation in the recommendations sent to the UK for an upcoming referendum was soundly rejected (Hiller, 1997b).

There was much concurrent discussion during this period between the Canadian and UK governments about their own preferred future for Newfoundland, of which the general population of the country was not aware (Hiller, 1997a; Neary, 1983,1988). But there was, among Newfoundlanders and Labradorians, a growing awareness of Canada and what Confederation could offer. This awareness was helped in no small way by Smallwood who, in response to his failed motion to have the Convention put forward Confederation as an option for the referendum ballot, struck a petition to advance his cause. With that option gaining so much public attention and with its own preference clearly a factor, the UK government was able to rationalize its inclusion on the referendum ballot.

When Clement Attlee visited Newfoundland in September 1942, in his capacities as Secretary of State for Dominion Affairs and Deputy Prime Minister, he had fresh to mind recent reports about the men of the NOFU. It had been a year of crisis for the unit during which Attlee's voice was instrumental in forging meaningful privileges for the NOFU men, privileges that recognized their needs and created increased parity with their Canadian counterparts. He had also visited some NOFU camps, as well as Newfoundland artillery units, just prior to his trip. In his public address to the people, he praised their contributions, adding that after the war, "Newfoundland like the rest of the world will be opening a new chapter in its history" (Attlee, 1942e, p. 7).

His account of the trip cast some light on the direction of his thinking about this new chapter—the future of Newfoundland—and the concerns he felt about what he observed: an aloof Commission of Government, no local governance structures, embryonic labour organization, little economic diversification, excessive financial control by a merchant class, a grossly disproportionate income gap between said merchant class and the rest of the people, and an absence of democratic practices.

With his background in social reform and experiences with economically marginalized communities, Attlee understood some of the historical reasons for what he saw, including the role his own country had played through controlling merchants "who in their own supposed interests prevented the growth of a settled community" (Attlee, 1942, as cited in Neary, 1984, p. 105). Taken together, these observations and insights likely deepened his feeling for Newfoundland and the direction it would take. He had expressed his concern in the 1933 debates about a commission of government for the country, which he did not support (Attlee, 1933). Attlee's visit also deepened his belief that political education was essential to the growth of democratic practice in Newfoundland, something he and his office had emphasized earlier in 1942 in discussions following the Goodwill Mission. The NOFU were its first beneficiaries.

When the UK government exerted its power to determine the options that were offered on the referendum about Newfoundland's political future, it did so with some hope that its (and Canada's) preferred choice—Confederation—would win approval. Knowing the unpopularity of the NCG—as Attlee wrote, "no one likes the Commission" (Attlee, 1942, as cited in Neary, 1984, p. 106)—the UK government also stipulated that a majority vote for one option was required. This requirement increased the likelihood that if a second referendum were needed, the option of extending the NCG would be dropped from the ballot. That is, in fact, what occurred. The second referendum's choices were: a return to responsible government or union with Canada.

The trepidation expressed by many about a return to responsible government—and the destitution and political chaos associated with it in its later years—when set against the reassurance of welfare-state security (features brought forth by Smallwood and others to advance the option of Confederation in the context of postwar economic decline), makes the majority vote for Confederation in the second referendum understandable. The harsh realities of the late era of responsible government that had preceded the NCG were brought to mind in the very way the option was stated on the referendum ballot: "Responsible Government, as it existed in 1933" (G. Long, 1999). And so, by a slim margin, a majority of Newfoundlanders and Labradorians voted practically—for a future they could imagine with hope, rather than for some version of one they had known and feared would come back to haunt them.[14]

It was compelling political drama. Thomas Curran would comment to those at a VAC Standing Committee hearing in 1960 about "how quickly we got Confederation in Newfoundland" (Canada, Parliament, House of Commons, Standing Committee on Veterans Affairs, 1960, p. 105). A political option that had been a blip on the public landscape in spring of 1946 was a legislative reality three years later.

There is no way to know how the NOFU, so many of whom had staunchly supported a return to responsible government, responded to the Confederation option being put forward while many of them were just returning from the UK.[15] The evidence of their consideration for their homeland while they were away, however, suggests a thoughtfulness about the situation and points to some collective will to impact it positively. The rural orientation of most of the NOFU members suggests many of them, like others in their communities, may have voted for Confederation with Canada. But there were also regional nuances that temper that speculation (Cadigan, 2009). Confederation—while not a return to self-government—offered a solid middle ground: becoming part of an interdependent federation from which Newfoundlanders and Labradorians could continue to negotiate a place and an identity in a postwar world, not the least part of which would be learning to be a Canadian citizen rather than a British subject, a transition they shared with all Canadians at that time.

In 1947, the Government of Canada had passed the Canadian Citizenship Act, which created the legal designation "Canadian citizen." Previously, Canadians, like Newfoundlanders and Labradorians, were British subjects,

---

14 For a recent and fulsome discussion of the decision to enter Confederation, see Blake & Baker (2019). The pro-Confederation majority was drawn largely from outside the Avalon Peninsula. In St. John's, the results were met with dismay. To mark the death of their country, many drew their house blinds and placed a letter D (for death) in their windows—at the time a common funeral practice at the home of a deceased person (Goss, 1980).

15 This endorsement was expressed in many letters to the editors of the *Bulletin* and through reports of camp committees of the NOFA. See, for example, Deisher camp's urging of the NOFA "to do everything in its power for a return to self-government in Newfoundland" ("Camp Meeting," 1945, p. 9).

that is, subjects of the monarch whom they served. The 1947 change was more than a semantic shuffle. Citizenship bestowed rights and protections; it underscored the responsibility of the state to the people it served and not the loyalty of the people to a head of state (Boyko, 2021). Not only were Newfoundlanders and Labradorians joining a new country, they were gaining a new status previously unknown to them: they would be citizens.

Shut Out Again: Negotiating the Terms of Union with Canada

Following the results of the second referendum on July 22, 1948, Newfoundland prepared to transition from country to province. Representatives were soon sent to Ottawa to negotiate the legal Terms of Union with Canada. As part of this process, the Chief Forestry Officer with the NCG's Department of Natural Resources and ex-NOFU Officer-in-Charge Jack Turner travelled to Ottawa in the early fall to discuss forestry issues and to highlight the case for benefits for the men he had led during the war (Baker & Miller Pitt, 1996). Turner died suddenly in Ottawa on September 27, however. The news reverberated throughout Newfoundland, most especially among those who had worked with him in the Department of Natural Resources and the thousands who had served with him in the UK.

Turner had been a strong, insistent advocate for his men; his death created a huge void in any advocacy for the ex-NOFU members during negotiations in Ottawa. Josh Goobie (NOFU #2308) described the impact for the unit as "the end of all things" (Goobie, n.d.). Without Turner there to speak for them, the men were orphaned under the final Terms of Union—as with the NCG's civil re-establishment plan, they were effectively shut out. No one else at the Department of Natural Resources was familiar with the work of the NOFU during the war or able and willing to advance its case in Ottawa.

The Terms of Union were thus finalized without NOFU representation—and also without any contributions from delegates or representatives of the Great War Veterans' Association; they represented the interests of Newfoundland and Labrador veterans and could have challenged the position that there was no Canadian legislation under which ex-NOFU might benefit (NOFA, 1949). It seemed that the NOFU's second shot at recognition had passed.

The memorandum of agreement that outlined the terms under which Newfoundland would become the tenth province of Canada was signed on December 11, 1948.[16] Term 38 of that agreement outlined the place of the province's veterans in their new country. The preface captured the spirit of the deal struck for Newfoundland's veterans: as part of Canada, they would receive the array of benefits available to Canadian veterans, "as if the Newfoundland veterans had served in His Majesty's Canadian forces" ("Terms of Union," 1948, p. 22).

In its brief to the Government of Canada, written shortly after Confederation, the NOFA challenged its exclusion from negotiations and the lack of consultation with its officials, insisting a delegation of NOFA officials should have accompanied the Great War Veterans' Association in negotiations around wartime service benefits, to speak on their behalf (NOFA, 1949). They argued that none of their representatives "were in possession of all the facts" about the NOFU, which only someone directly connected to the unit would know (NOFA, 1949). This argument suggested that a lack of insight about the work of the NOFU during the war may have shaped the outcome of negotiations. It also underscored that the work of the unit as an essential wartime service was neither understood nor valued—an essential point that would be made over and over in arguments in favour of its recognition, arguments that would now have to take place within the framework of Canadian democracy.

---

16 Many aspects of the final Terms of Union were hotly contested in the years following Confederation. Among the more egregious and devastating omissions from the agreed terms were the recognition and rights of the Indigenous Peoples of the Island of Newfoundland and mainland Labrador.

Veterans Affairs Canada: Another Struggle for Recognition[17]

The ex-NOFU members entered Confederation with both unfinished business and more reservations about their place in the new country than their compatriots in other services, whose concessions had been well and fully negotiated by the Great War Veterans' Association (Neary, 2011b). They had already suffered two major setbacks: NOFU members had been shut out of most concessions offered by the NCG in civil re-establishment and their concerns were inadequately represented in the negotiation of the Terms of Union with Canada. Early in Confederation, they were set on their heels again by the rejection of their efforts to have the Government of Canada recognize them as veterans. The stage was then set for a prolonged lesson in democratic citizenship—that its rights and rewards were not necessarily givens; they were—and, for many, would remain—to be secured through struggle. Such was the case for veterans from three categories of service in postwar Newfoundland and early Confederation Newfoundland: the Merchant Navy, the Rescue Tug Service, and the NOFU.

Interestingly, Joseph Smallwood's base of support for both Confederation and his election as the province's first premier was in rural Newfoundland, where most of the ex-NOFU men lived. Yet the Premier did little to advance their rights after the union with Canada. Smallwood was made aware of the difficulties the men faced, time and again; many of them wrote seeking help in attaining support in a range of issues related to re-establishment, from requests for help in securing allowances and benefits to aid in acquiring land.[18] These letters were forwarded from the Premier's Office to the federal Department of Veterans Affairs (VAC) office in St. John's, which sent a reply of explanation (and sometimes exasperation, if a letter writer had already approached the office several times).[19] As the VAC officers reminded Smallwood in a pro forma

---

17 See also Neary (1993b) for an account of the NOFA efforts to win concessions for ex-NOFU men.

18 See, for example, Ryan (1949) and March (1964).

19 See numerous examples of such letters in the J. R. Smallwood papers, Collection 075, File 3.10.031, Archives and Special Collections, Memorial University.

response each time: "Ex-Members of the Newfoundland Forestry Unit were not considered for rehabilitation benefits in any way except in the case where they were interested in farming operations, where they could obtain assistance under Plan B" (Vivian, 1949). In their eyes, the matter ended there.

In 1957, John Diefenbaker had replaced Louis St. Laurent as Canada's prime minister. Hopeful that a change of leadership might offer opportunity, the NOFA prepared to re-state its case, applying for a hearing at the VAC Parliamentary Standing Committee. It took place in 1960. Letters exchanged between Premier Smallwood and VAC Minister Milton Gregg in July 1950, a little over a year after Confederation, were not tabled at this hearing nor attached to the minutes, although VAC Minister Alfred J. Brooks had requested same (Brooks, 1960). Conservative Member of Parliament James McGrath later suggested Smallwood had refused permission to release the correspondence (McGrath, 1962). Both the NOFA representatives and the Standing Committee members believed that these letters could provide important context for the exclusion of the NOFU from concessions for veterans.

At the hearing, Thomas Curran reported that Smallwood was largely oblivious of the plight of the NOFU men (Canada, Parliament, House of Commons, Standing Committee on Veterans Affairs, 1960; Curran, 1987)—and the premier admitted as much in the first of the series of letters he exchanged with VAC Minister Milton Gregg in July 1950. That letter had been drafted by the NOFA lawyer and flagged to Smallwood as being "of the type which we [the NOFA] would like to have you write [to VAC]" (Mercer, 1950b, p. 2). The association had not wanted the federal government to make a decision about their appeal in the absence of the Premier's input and support. The letter was sent to Smallwood on July 3, shortly after a brief outlining the NOFA case with the federal government and a request for support had been conveyed to him.

Smallwood signed the letter verbatim, and in so doing confirmed that "during the pre-Confederation talks and the later discussions on terms of union . . . I was not in possession of all the material facts" (Smallwood, 1950, p. 1). Further, "having [now] read the brief . . . if all the facts were then before us there is no doubt but that stronger arguments could and would have been put forward for recognition of the unit" (Smallwood, 1950, p. 2). It was low-grade support with no additional efforts made on behalf of the Premier for the NOFA case.

By signing the letter, Smallwood also supported the NOFA request for "some form of rehabilitation benefits," hospital and medical benefits for illness or injury incurred during employment, and pensions where disability or death resulted from work injury (Smallwood, 1950). Gregg replied on July 7, 1950, noting reference in the Smallwood letter to a meeting of the Deputy Minister of VAC, Walter Woods, and NOFA executive and their solicitor, Isaac Mercer (of the St. John's law firm Mercer, Mifflin, O'Dea), which had taken place in St. John's in mid-June that year. Gregg explained that, at the meeting, the NOFA representatives had raised two issues: the first concerned recognizing the NOFU members as veterans, so they might receive preference in employment through the National Employment Service; the second requested recognizing them under the Civilian War Pensions and Allowances Act (Gregg, 1950a). He indicated that these matters were under investigation, and that he would inform Smallwood of the results.

Gregg wrote again on July 20, this time regarding recognizing NOFU members under the National Employment Act. Gregg commented that such recognition would create "considerable dilution of the veteran body," pointing out that several other groups, including merchant seamen, civilian overseas firefighters, and armed forces personnel without overseas experience, had no privileges under the Act (Gregg, 1950b, p. 1). He argued that "granting of Veterans' Preference to ex-members of the Forestry Unit would, therefore, put them in a preferred class of civilian war-worker . . . [and] it would not appear as though the request could be entertained" (Gregg, 1950b, p. 1). The comment harkened back to the UK government's responses to requests from the NOFU, requests that were seen as an effort to achieve preferential rather than similar treatment.

In a subsequent letter written July 25, Gregg challenged the common assertion that the full facts about the unit had not been discussed in pre-Confederation talks. The Deputy Minister reminded the Premier that the case of the NOFU had been discussed in the proposal-for-union stage of the Confederation conversation, and at the Terms of Union discussions that followed the referendum (Gregg, 1950c). He insisted that on both occasions the conclusion was that there was "no enabling Canadian authority by which [the ex-NOFU members] could become eligible" for benefits. Nor was it considered that "the circumstances of their occupation were such as to warrant special consideration" (Gregg, 1950c, p. 1).

Gregg also reminded the Premier that the NCG had accepted no responsibility for the ex-NOFU, therefore there was no liability for the Government of Canada to carry over, as there had been with the other pension liabilities of both the First and Second World Wars. For Gregg, the terms of the contract signed by the NOFU workers were clear about illness or accident—those affected were covered by the UK Workmen's Compensation Act. Gregg felt that no compelling argument had been made, and he showed no inclination to move beyond the strict terms of the existing Civilian War Pensions and Allowances Act (Gregg, 1950c). A letter to Isaac Mercer from VAC Deputy Minister E. L. M. Burns dated August 4, 1950, was curt, quoting large chunks from the Gregg letter to Smallwood. Burns assured Mercer that under Minister Brooks the VAC had "examined carefully and given every consideration" to the requests of his client (the NOFA), but "favourable consideration to these requests was not possible" (Burns, 1950, p. 1). The letter was clearly intended to put the matter to rest.

But the NOFA was not dissuaded. It continued its efforts and finally, on March 31, 1960—some fifteen years after the end of the war and over a decade after Confederation—President Thomas Curran and Secretary Charles R. Baggs appeared for the first time before the Parliamentary Standing Committee on Veterans Affairs at Ottawa. They presented a brief outlining the case for recognizing the NOFU. The meeting was the first real attention the organization had won; in Curran's words, all the previous attempts by Isaac Mercer—whether through correspondence or meetings with the VAC—had "got nowhere . . . not even to first base" (Canada, Parliament, House of Commons, Standing Committee on Veterans Affairs, 1960, p. 113).

In February 1960, just over a month prior to that VAC hearing, the first branch of the NOFA was formed at Placentia. At its inaugural meeting, several guests were present, including a local magistrate, mayor, and doctor, as well as Esau Thoms, a founding member of the provincial New Democratic Party and president of the Newfoundland Federation of Labour (1960–68). Thoms's presence and pledge of support positioned the NOFA efforts as a labour struggle centred on the democratic rights of workers in Canada, as well as one for recognition of wartime service in a capacity other than armed combat. Thoms had drafted a resolution for the new NOFA branch, which was approved at that first meeting. By this time, the priority of the NOFA had developed into securing parity of benefits with the Canadian Forestry Corps. Thoms's resolution focussed on that point, requesting that the Government of Canada set in motion the necessary steps to achieve it ("Placentia," 1960).

With more battles to fight to gain such parity, the NOFA continued efforts to strengthen its presence at home. Following the 1960 VAC hearing, a St. John's branch of the NOFA was formed, again with the support of Esau Thoms ("Local Foresters," 1960). With this development and their VAC committee appearance, Curran and Baggs were feeling some wind in their sails in terms of organizational momentum. The NOFA had understandably been in a period of transition after the war, as members adjusted to repatriation without support while seeking employment in and reintegration into a country preoccupied with a National Convention and two referenda. When that political dust settled, the NOFA had to remount its efforts while adjusting to a new country with a new system of governance.

In 1962, Curran estimated the membership of the NOFA at approximately a thousand men, not even a third of the total unit numbers (Canada, Parliament, House of Commons, Standing Committee on Veterans Affairs, 1960). In 1946, as the work of the unit was coming to an end in Scotland, the executive had stressed the importance of maintaining and gaining membership once members had repatriated, fully realizing that robust branches were needed to handle the business of the association. They foresaw that, in the face of a near-complete absence of government support, the NOFA's work would include providing information about employment opportunities and offering support for members flailing in their efforts to re-establish a life in Newfoundland. A solid organization on home turf could lobby for members and throw its weight "behind every movement for the good of our country" ("The Foresters Return," 1946, p. 2).

This membership surge did not quickly materialize, however. Geographic dispersal and reintegration preoccupations aside, limited membership uptake raises questions about how some ex-NOFU members may have seen the push for benefits—or even what they thought of the NOFA itself. There had been holdouts while in the UK, but the seemingly muted interest in the organization at home (as indicated by membership numbers) was incongruous with what was at stake—allowances, pensions, and other benefits. It suggests the impact of re-entry and the public attitudes toward the unit that were part of it.

At the end of the NOFU members' time in the UK, NOFA Secretary Rex Browne had cast the organization to local press "not as a union, but a brotherhood" ("Foresters Association to Function," 1946, p. 1). Five years following the 1960 NOFA presentation to the VAC, the "brotherhood" had grown to an estimated 2,300 members and included many who had served in the NOFU only a short time before transferring to one of the armed forces (Day, 1965). The association had finally found its legs on the home front—as evidenced by increased membership, branch-level organization, and lobbying with letter-writing and appeals—but only after it got and held the attention of the Government of Canada, a decade-and-a-half after repatriation.

A Foundational Document

A *Brief Submitted on Behalf of the Newfoundland Overseas Forestry Unit* set out in fourteen pages the case for recognition of the wartime service of the ex-NOFU under the Veterans Charter (a suite of Canadian postwar programs and benefits). Purportedly prepared in 1949 by NOFA lawyer Isaac Mercer, it was based on input from Austin Baird, Thomas Curran, and other members of the NOFA executive.[20] A copy of the brief was sent to Premier Smallwood on April 29, 1950, along with a request for his support. The brief was also presented to the provincial command of the Canadian Legion that same year (Matthews, 1961b).

At the VAC hearing in 1960, Curran presented the brief, along with a resolution of support from the Placentia branch of the NOFA. He fielded

20 It was also speculated that the brief—perhaps an earlier version—was part of the materials Jack Turner had prepared for the 1948 discussions in Ottawa, in the lead up to the drafting of the Terms of Union (Matthews, 1963).

evidentiary questions from committee members during the meeting, acknowledging that the brief was unchanged from the 1949 original.

The brief itself is a sprawling account of the ways in which the men of the unit felt they had been discriminated against and "given no equality of preference for employment; no rehabilitation benefits; no pension benefits; no sick or hospital benefits and in fact no recognition in any way" (NOFA, 1949, p. 1)—despite having done what they considered to be the same work as the Newfoundland Forestry Companies of the First World War and the Canadian Forestry Corps, both of which, as military units, were included in the Veterans Charter. The evidentiary documentation for the brief had been gathered in the years following repatriation; it ranged from excerpts of official correspondence and commendations to rebuttals of arguments against the NOFU case circulating at the time.

The brief was an appeal for inclusion under two specific pieces of legislation: the Civilian War Pensions and Allowances Act and the Special Operators War Service Benefits Act (NOFA, 1949). An argument for inclusion under the Veterans Charter was not directly advanced in the brief; this was perhaps because of the earlier failures on that front—to be included as on par with veterans—that the NOFU and NOFA had experienced with the NCG and VAC. Nevertheless, this decision defined the lengths to which Members of Parliament were willing to go to include the unit in existing legislation and, at the same time, to be seen as addressing its demands (McGrath, 1962).

The NOFA presentation to VAC was effective in that it mobilized support among parliamentarians and prompted a review of their case and subsequent changes to the Civilian War Pensions and Allowances Act under Bill C-64. In Part XI of that Act, benefits were extended to certain civilian groups who gave meritorious service in support of, but not part of, the armed forces. The amendment included the NOFU, giving eligibility for war allowances only (not pensions or disability allowances) to some of its members—those who had signed for the duration of the war and who had served a minimum of six months in the UK—and it included benefits for wives and children of eligible members who had died (Churchill, 1962).

These advances aside, that the ex-NOFU members were still categorized as civilians and not included in the Veterans Charter remained a contentious point for some (C. Carter, 1962; Matthews, 1962, 1963; "Placentia Branch Foresters," 1962). As well, the "signed for the duration" clause was considered very limiting, because many men had served for the duration without signing for the duration, renewing every six months instead (Canada, Parliament, House of Commons, Standing Committee on Veterans Affairs, 1960). Civil re-establishment, educational studies, and other benefits still eluded the ex-NOFU.

The Placentia branch of the NOFA, mostly through the letter-writing campaign of its chair, Augustus Greene (NOFU #2255), made a high-profile contribution to the debate around ex-NOFU rights. Following the decision to amend the Civilian War Pensions and Allowances Act in 1962 to include certain NOFU members, the Branch once more raised its voice. Greene penned letters to Chesley Carter, then the Member for the riding of Burin-Burgeo, as well as VAC Minister Gordon Churchill and Prime Minister Diefenbaker.[21] From the point of view of Government of Canada officials, ex-NOFU men simply did not qualify as veterans and were therefore not eligible for benefits under the War Pensions Act—they had been civilian labourers doing wartime work. As Churchill said in his reply to Augustus Greene: "The basic definition of a 'veteran' is a former member of HM armed forces and consequently cannot apply to former members of the Newfoundland Forestry Unit of World War 2 or to the other civilian groups covered under the CWPAA [Civilian War Pensions and Allowances Act]" (Churchill, 1962, p. 2709).

Similar views circulated among the public and within the NOFU itself. Newspapers carried fiery exchanges among veterans who refuted the legitimacy of the ex-NOFU members' and their supporters' claims for on-par status. Wickford Collins, a veteran of the Royal Newfoundland Artillery (# 970849) wrote in a letter to the editor of *The Daily News*:

> It is not doubted that along with miners, factory workers and farmers they were a significant part of the civilian war effort and there is possibly considerable merit in their claim that their

---

21 Esau Thoms sent Greene's letter to the Prime Minister, copied to leaders of all federal parties, as well as to *The Evening Telegram*, where it was subsequently published (Greene, 1962).

> efforts have been poorly rewarded, but their negotiations with the Government of Canada should come under the Civilian Allowances Act and not by any stretch of the imagination under the Veterans Charter. The right to be called a veteran is earned at a price the foresters most certainly did not pay and no government, no parliament can pay it for them now by legislative action. (W. Collins, 1963, p. 4)

Augustus Greene penned an angry reply to Collins (Greene, 1963) and others who wrote to make a similar point (Greene, 1965), including to one veteran who wrote to say, "Forget it! . . . After twenty years, it's time to give up" (War Veteran, 1965, p. 6). There was some quiet dissent within the unit, as well. Ex-NOFU and Home Guard member John Nick Jeddore admitted to feeling uncomfortable about being called a veteran: "I did get a medal, and was entitled to one more, but I didn't bother to apply. I wasn't a soldier, sailor, or airman. I cut down trees, drove a team, checked timber, and did a few other jobs. The one medal I have, I keep hidden" (Jeddore, 2015, p. 117). His comment raises a question about the extent to which disagreement about, or discomfort with, efforts to acquire veteran status for the ex-NOFU existed within as well as outside the unit.

But many people, including some Members of Parliament and members of the Royal Canadian Legion, felt the men had not received a just settlement from the Government of Canada. Among the more vocal Members of Parliament was Chesley Carter, a veteran himself, who was adamant that the NOFU members were dealt an injustice in 1962. In a letter to Augustus Greene, he asserted the amendments to the Civilian War Pensions and Allowances Act constituted "a different course—and then taken deliberately—to make certain that [the ex-NOFU men] would not be included in the Veterans Charter" ("Charges Foresters," 1963, p. 3).

The struggle to gain more and better benefits for more of the ex-NOFU men than were covered by the Bill C-64 amendments continued through the early to mid-1960s. Several Members—Carter, Herman Batten, James Tucker, Charles Granger, and John Pickersgill—kept the issue alive in Canada's parliament. Late in the fall of 1962, the VAC announced that the "signed for the duration" clause had been removed from the legislation and replaced with an eligibility requirement including ex-members who had served until the end of the war—that is, those who were discharged after May 8, 1945 (Churchill, 1962).

The NOFA also continued to lobby strongly for fuller recognition, now with increased membership and a wider branch system. Additional politicians came onside, including some of the province's Members of the House of Assembly, who appreciated not only the righteousness of the cause but also the large group of residents—potential voters—across all districts whose lives were touched by the issue, estimated to be between ten and twelve thousand people when members and their families were counted ("Foresters Still Pressing," 1964). One such Member, Progressive Conservative William Browne, recommended the NOFA prepare another brief, stating their case more strongly as it had evolved over the years ("Foresters Seek Benefits," 1963).

Slowly but surely, as they persisted in their efforts, progress was achieved. By 1967, the Civilian War Pensions and Allowances Act was further amended to include those who had served "for a period of more than six months" or "for any period, if the service was terminated on medical grounds" (*Bill C-180*, 1967). By then, more than twenty years after repatriation, the ex-NOFU, as Peter Neary put it, "had slowly clawed their way into the [VAC] system" (Neary, 2011b, p. 286). In 1991, a Subcommittee on Veterans Affairs, in a report aptly entitled *It's Almost Too Late*, recommended that the Veterans Independence Pension, which assisted eligible applicants to remain in their homes as they aged, be extended to all who were covered in Part XI of the Civilian War Pensions and Allowances Act (Canada, Parliament, Senate, Subcommittee on Veterans Affairs, 1991).

Still clawing in 2000, the ex-NOFU of the Second World War were finally included "as though" veterans of a theatre of actual war by virtue of Bill C-41—an act to amend the statute law in relation to veteran's benefits.[22] The amendment (Part II.1) to the Pension Act read: "The Pension Act applies to and in respect of members of the Newfoundland Overseas Forestry Unit

22 This criterion, as outlined in the Civilian War-Related Benefits Act, remained a point of exclusion after 2000. See "Foresters Waiting" (2003).

as though their service as such during the War had been service as a member of the forces within the meaning of that Act" (*Bill* C-*41*, 1999, Clause 19.1). The legislation was passed just a month shy of the sixtieth anniversary of the November 1939 formation of the unit.[23] By then, the youngest ex-NOFU men, who had joined up at the onset of the war at the minimum age of 18, were 80 years old, and those who had entered at the maximum age of eligibility—50—were long deceased.[24] In his remarks to Parliament on the second reading of the bill, Newfoundland Member of Parliament George Baker, Minister of Veterans Affairs, commented that the bill "does something that should have been done many years ago" and, with specific reference to the ex-NOFU men, of whom he estimated only five hundred to a thousand remained, he remarked that what the bill accomplished "should have been in the Terms of Union" (G. S. Baker, 2000).

As a new century dawned, the scope of veterans' benefits in Canada had broadened profoundly—and the persistent efforts of the NOFA on behalf of the men of the NOFU had been a part of that evolution. It would continue to broaden, as the landscape of war and service and the tolls they exact changed, leading to the adoption of the New Veterans Charter in 2005.

The Royal Canadian Legion and the NOFU

In the post-Second World War period, the Royal Canadian Legion was a strong lobbying force for veterans of the armed forces. Its support was considered central to the advancement of ex-NOFU member rights and the unit's recognition by the Government of Canada. Aside from its consideration of the righteousness of the NOFU cause, there were also First World War veterans—who were Legion members—who had joined the NOFU and the Merchant Navy during the Second World War (Great War Veterans'

23 The bill also extended benefits to other civilian groups who served overseas, including the Voluntary Aid Detachment of the First World War, Ferry Command air pilots, welfare workers, and the Corps of Canadian Firefighters for Service in the UK.

24 Many who led the challenges to the exclusive legislation, including Thomas Curran, had died by this time. In his comments to the House, Baker mentioned the efforts of Lloyd Thompson (NOFU #1328) of Botwood, who had been a camp clerk with the unit and who participated in the negotiations on behalf of the ex-NOFU men (G. S. Baker, 2000).

Association, 1940) and to whom the Legion owed some responsibility. The efforts to gain the Legion's recognition, however, was not straightforward and ran parallel to NOFA efforts on the Government of Canada front.

In 1950, Isaac Mercer presented several copies of the 1949 NOFA brief to the Legion provincial command, in preparation for the Dominion (Canadian) Command convention in Winnipeg that year. According to Mercer, it was the hope of the NOFA to secure the support of the Canadian command to strengthen their position with the federal government (Mercer, 1950a). In this regard, the Provincial Command had passed the following resolution:

> Resolved that this convention approves of the brief submitted by the Newfoundland Overseas Forestry Unit asking for consideration of the federal government in respect of rehabilitation pensions, hospitalization, and a degree of preference in employment in the Civil Service, rating next below ex-service, and that the Provincial Council be directed to ask Dominion Command to place the matter before the Federal Government for further consideration. ("Legion's Provincial Command," 1950, p. 1)

But just as the NOFA leadership was new to and unfamiliar with the workings of Canadian bureaucracy, the Newfoundland legionnaires were new to the Dominion Command of the Legion and its workings. Apparently, the resolution was not presented to Dominion Command in time for the proper procedures to unfold around adding a resolution to its convention program. Delegates did not have sufficient opportunity to study the case; therefore, the resolution could not be debated on the convention floor (Matthews, 1962). According to Frederick Matthews, a legionnaire who attended the convention, there was also no record of the brief ever reaching the national Legion executive until much later in the decade (Matthews, 1965a, 1965b).[25]

---

25 Some believed there was something more nefarious about the disappearance of the brief and the missed opportunity to present the resolution to Dominion Command, citing some Legion members' negative attitudes about the status of the ex-NOFU and suggesting an inner scuttling of their case. See Crews (1965).

Matthews wrote a regular column on veterans' issues in *The Western Star* and often shared context for many of the service issues that came up in letters he received. In one of the many columns he penned about the ex-NOFU situation, Matthews insisted that throughout the 1950s, the Provincial Command continued to support recognizing the men (Matthews, 1961a), including through pensions for those who were disabled or suffered impaired health while at work in the UK (Marshall, 1951). But beyond the failed efforts to promote their cause in 1950 in Winnipeg, the issue seemed to sit in a Legion limbo similar to the one it was in with the Government of Canada. It was not until 1958 that the Provincial Command formally endorsed the NOFA efforts in a letter of support, as they prepared for their VAC hearing ("Legion to Back Foresters," 1958). This endorsement was significant to members of the Standing Committee when Curran shared it early in his presentation (Canada, Parliament, House of Commons, Standing Committee on Veterans Affairs, 1960).

A few months after the 1960 VAC hearing, the Canadian Legion determined that some ex-NOFU members were eligible for ordinary (voting) membership in the Legion, at the discretion of branches, if they had signed for the duration and met two conditions: service in the UK-based Home Guard, and documented receipt of the Defence Medal (Matthews, 1961a). Under these conditions, approximately 20 percent of ex-NOFU members became eligible for membership in the Legion. The two-tiered division of the NOFU—those who were uniformed and had military training through the Home Guard and the rest of the members—was enshrined in Legion by-laws.

Further division was created when some provincial Legion branches exercised their discretion and voted to disallow membership to ex-NOFU men, regardless of their qualification under Provincial Command guidelines. This discretion resulted in membership eligibility having a peculiar geographic arbitrariness. The behaviour of a few branches aside, the Dominion Command also sent a brief to the Prime Minister and federal cabinet in 1960 (Matthews, 1961b) expressing support for ex-NOFU members.

The uneven treatment continued. In 1972, when more ex-NOFU members became eligible for Legion membership, those who had not been awarded the Defence Medal were offered associate (non-voting, later called affiliate) rather than full membership (Royal Canadian Legion, 2016). To this day, debates about membership hierarchies and focus within the Legion continue (Bruyea & Smol, 2016).

The Volunteer Service Medal

The fight for recognition emerged again in 1981, albeit it on a provincial scale, when the Government of Newfoundland passed The Volunteer War Service Medal Act. In what seemed a move of utter unawareness of or indifference to over three decades of NOFA efforts, the terms of the Act fell back on the same language and definition of "veteran" that had excluded ex-NOFU men for thirty-five years. The Act was intended to recognize those from the former Dominion of Newfoundland who had served in the UK armed forces but who had not received a volunteer war service medal from another country (including Canada). According to the Act, a veteran was a person who "volunteered to serve in the British Imperial Forces during the Second World War" (*Volunteer War Services Medal*, 1981, p. 258). The first medals were awarded at Remembrance Day ceremonies in St. John's on November 11, 1981.

At the annual meeting of the NOFA held in Corner Brook in September 1982, members revealed that they had failed to get the backing of the Legion in their request for eligibility for the medal. An appeal to the provincial government met with an offer to strike a special medal for the ex-NOFU, a proposal which the NOFA rejected as "belittling," given that no other service unit had a distinct medal ("Foresters Association Wants," 1982, p. 5). By 1983, the Legion had received from its membership the authority to pursue an amendment to the medal legislation to include the ex-NOFU men ("Banquet," 1983).

In the glacial movement forward on the path of recognition that had become a hallmark of the ex-NOFU endeavours, ten years later the legislation was amended, this time to include those who had been excluded at Confederation. The new act named as eligible all those

> who volunteered and served in units or organizations raised or maintained as the contribution of the Dominion of Newfoundland to the allied war effort, namely, the British Imperial Forces,

the Newfoundland Overseas Forestry Unit, the merchant navy and the Newfoundland Regiment, or, a unit or organization prescribed by the Lieutenant-Governor in Council." (*Volunteer War Services Medal Act*, 1993, clause 3a)

Postwar Forestry Sector Unrest: The Newfoundland Loggers' Strike of 1959

In the years following Confederation, postwar recognition for their overseas work was not the only issue that preoccupied many ex-NOFU members. Through their work as loggers prior to the war, many of the men had directly experienced the gross inequities in their industry that Gordon Bradley had documented in 1934 (Bradley, 1934, as cited in Neary, 1985), which continued to shape the woods industry throughout the first half of the twentieth century. These inequities were, in part, what had motivated some of the men to join the NOFU in the first place—they traded poor work conditions at home for the higher calling of overseas work to support the Allied war effort. With the return to employment at one of the two large pulp and paper companies after the war, many men found themselves once again in the throes of a struggle for the rights of woods workers. The fight for better wages and working conditions ran parallel to the ex-NOFU members' efforts for postwar benefits. Both were critical moments in the history of woods work in twentieth-century Newfoundland and Labrador, although the dramatic labour strife among woods workers in the 1950s led to the greatest notoriety.

By the onset of the war, the Newfoundland Lumbermen's Association, under the leadership of Joseph Thompson, had lost its status as the only organization for woods workers. In 1938, splinter organizations had developed: the Newfoundland Labourers' Union, led by Percy Fudge, formed at Corner Brook, and the Workers' Central Protective Union, led by Charles Tulk, formed at Deer Lake. In 1940, the NCG created the Newfoundland Woods Labour Board, which brought together representatives from the companies and the unions to address and resolve issues, in order to ensure stability in the industry during the war. From the 1940s until the late 1950s, there were no work stoppages, a marked contrast to the woods labour unrest that had marked the 1930s (Sutherland, 1992).

By the mid-1950s, loggers' disaffection with the Newfoundland Lumbermen's Association and its leader, Joseph J. Thompson, as well as with the other unions, had grown. The International Woodworkers of America moved into British Columbia in the 1940s, successfully organizing and striking for improved benefits for woods workers there. In 1956, frustrated loggers invited the union to Newfoundland. Following a concerted drive, loggers gave it and its Eastern Director, H. Landon Ladd, an overwhelming endorsement. Ladd attempted unsuccessfully to negotiate a collective agreement with the Anglo-Newfoundland Development Company. On December 31, 1958, after the ANDCO refused the report of a conciliation board, loggers withdrew their labour, closing camps and curtailing production for several weeks (Seymour, 1976).

A large group of loggers rallies to listen to the IWA's H. Landon Ladd, ca. 1958. (Courtesy of William Ready Division of Archives and Research Collections, McMaster University, Edward E. Seymour Fonds, 35–8-72)

Seeing Ladd's leadership charisma and influence with loggers as a threat to his own power, Premier Smallwood vilified the union. In an address to the province on February 12, 1959, Smallwood reminded the largely rural-based loggers of their role in securing Confederation, by then nearing its tenth anniversary, and ensuring its future success. He also stoked xenophobia, casting Ladd and the union as outsiders who had no place among "decent, God-fearing people" who were "made dirty by the presence of this wicked and mischievous body of reckless and irresponsible wreckers" (Smallwood, 1959, as cited in Paine, 1986, p. 205). His radio rhetoric mobilized a large base of antagonism for the union and less than a month later, Smallwood decertified it. Public opinion against the union was sealed when, during a mêlée between striking loggers and police at Badger on March 10—the Badger Riot, the nadir of what became known as the Newfoundland Loggers' Strike—a young Newfoundland Constabulary Constable, William Moss, was struck on the head and died of his injury. On March 12, the ANDCO signed an agreement with the government-approved replacement union, the Newfoundland Brotherhood of Wood Workers, that featured many of the same concessions Ladd had sought. By 1961, the United Brotherhood of Carpenters and Joiners of America officially represented loggers.[26]

The strike was a tragic, unsettling culmination to a half-century of struggle by woods workers against unfair attitudes, treatment, and policy. It divided the province and set public opinion in the rest of Canada against the Smallwood government. In 1960, yet another inquiry into the logging industry was struck. Led by Brian Dunfield, this Royal Commission of Enquiry confirmed what loggers had been saying and experiencing for decades. Dunfield recommended many of the changes for which the loggers had struck in 1959.[27]

Despite his early-career socialist leanings, the rights of workers, especially woods workers, was not a Smallwood priority, as his callous manipulation of public sentiment and his heavy-handed treatment of the loggers during the

26 For a detailed account of the strike, see P. W. Collins (2014).

27 See Newfoundland Commission of Enquiry (1960).

strike had demonstrated (Cadigan, 2006; Neary, 1969). Smallwood's Liberal Party relied on the financial backing of the pulp and paper companies, and it was in their interests that he acted (Power, as cited in Porter, 1994). For a new province, it was a harsh and divisive time that played out on the national stage; it included what Peter Neary called "a use of state power which would have been considered morally indefensible almost anywhere else in Canada" (Neary, 1969, p. 41). For loggers, it was another reminder of their place in a social hierarchy that was based on exploited labour and rural location. They were used as political pawns when convenient, but their own needs were otherwise largely ignored. These prevailing attitudes created a charged context in which to make a case for benefits for the overseas wartime loggers.

The unrest in the woods industry during the 1950s impacted the NOFA efforts to pursue their case with the Government of Canada. During his testimony at the Standing Committee of the VAC in 1960, Thomas Curran referenced a year's delay in presenting their case by noting "the unfortunate circumstances in Newfoundland last year with the loggers' strike and, I being in the midst of it, we thought it better to layover" (Canada, Parliament, House of Commons, Standing Committee on Veterans Affairs, 1960, p. 104). Curran was perhaps also seeking to avert any thinking that the delay was linked to the friction then existing between Liberal Premier Smallwood and Progressive Conservative Prime Minister Diefenbaker, who had clashed over the premier's request for federal Royal Canadian Mounted Police reinforcements to quell unrest during the Loggers' Strike. Curran sought to distance the NOFA from this tension, adding that the delay had "nothing whatever to do with politics so far as I am concerned," and insisting that the ex-NOFU members' fight for recognition was "not a political issue" (Canada, Parliament, House of Commons, Standing Committee on Veterans Affairs, 1960, p. 105). It was, rather, a fight for rights and the benefits of citizenship, regardless of government stripe or the temperature of provincial-federal relations.[28]

---

28 For an account of the tensions between Smallwood and Diefenbaker, see J. J. S. Garner in Neary (2017).

The union-led Newfoundland Loggers' Strike is widely recognized as a key moment in the history of woods workers' struggles for rights; the NOFA's push for benefits constitutes a less dramatic and lesser known—but also meaningful—campaign for rights and recognition within the same sector. While the context out of which these struggles arose had their roots in prewar Newfoundland, overseas wartime experience separated them, making the roads to recognition and the tactics available to and used by each "brotherhood," as well as their outcomes, different. Their respective struggles thus contributed to increasingly differentiated identities—as loggers and foresters—within the broader spectrum of woods labour in the postwar, early Confederation context of Newfoundland and Labrador.

CONCLUSION

WORK, WARTIME SERVICE, AND REMEMBRANCE

The Newfoundland Overseas Forestry Unit of the Second World War was a large group of migrant workers recruited from a persistently undervalued occupational sector. They volunteered for wartime work under contract, as duly constituted by the Newfoundland Commission of Government and at the request of the United Kingdom government. The men travelled overseas in the early years of the war and many stayed with the unit not only for the duration of the conflict but for an additional year after it ended. The story of the unit—its status and service—is set within the fraught socio-cultural context of Newfoundland in the era of the Commission of Government, and is imbued with intergovernmental politics and colonial attitudes that shaped the unit's formation, its overseas experiences, and its postwar difficulties during a watershed time—Newfoundland's advent as a province of Canada. Their story has implications for expanded historical understanding in several intersecting areas, among them the value and validation of labour, the significance of occupational nomenclature and identity, the meaning of wartime service, and the nature of remembrance and commemoration.

Value, Status, and Recognition

Labour status and validation were key preoccupations of the unit throughout its existence, but most especially during the first three of its six and one-half years in the UK. Those first years were a time of great instability for the NOFU, as evidenced by declining worker numbers, inefficient productivity, and unsatisfactory member well-being and unit status. During this period, the reputation of the unit as well as its work and status were at stake, a situation exacerbated by unhelpful public relations resulting from poor discipline and conduct, which in turn often occurred in response to subpar conditions and treatment. Despite the fact that so many of the factors affecting its reputation were beyond the control of both the workers and the unit's Officer-in-Charge, the NOFU nonetheless wore the problems.

In 1942, the situation for the NOFU reached a turning point. During that time, a reset for the unit was accomplished due in large part to the careful calculations of unit leadership (including its superintendents), who demanded adherence to the legislative authority of the Act by which the unit had been formed and insisted on measures to improve the well-being of its men in the hopes that, given the dire labour needs of the UK, they would earn some concessions. These demands were facilitated by key players in the Dominions Office, including Clement Attlee, whose background and Labour affiliation predisposed him to an appreciation of workers' needs generally and the problems of their homeland, specifically. The Dominions Office understood what was at stake in the treatment of the NOFU and prioritized the well-being of the workers, the integrity of unit leadership, and the autonomy of the unit as the debate about the future of Newfoundland, in the context of its wartime economic turnaround, intensified. While the NCG was largely indifferent to the well-being of the unit, ongoing ill treatment and its results had the potential to increase animosity toward it and add hostility toward the UK government to the mix.

The Dominions Office earned the trust of the NOFU by challenging the UK's Ministry of Supply on its inequitable treatment of and disquieting attitudes toward the unit.[1] Since they were foreign guests assigned to what was by all measures the provision of essential wartime work, the NOFU members should have garnered more regard and value. Instead, the men were left to

1 The Dominions Office developed a reputation for such "skilful diplomacy," especially during the Second World War, and for maintaining what was a sometimes-shaky alliance of the dominions (A. Stewart, 2010, p. 43).

fend for themselves in often very isolated contexts, without due consideration for their overall well-being. That it took two years for discontentment to crest—and longer for the Ministry of Supply to respond—is indicative of the workers' restraint; most were focussed, instead, on the job for which they had come to the UK. While the Ministry of Supply shrieked about the large numbers who returned to Newfoundland or transferred to the armed forces when their contracts were up, it is remarkable that these numbers were not higher; that is, the question of why so many left the unit is more aptly supplanted by one of how so many remained, given their treatment.

The actions of the Dominions Office were not altruistic. In pursuing better conditions for the NOFU, it was in a calculated bid to maintain good relations with the people of Newfoundland as the UK government banked on divesting itself of future responsibilities for the troublesome colony. During the welfare crisis of 1942, the Dominions Office was able to promote its agenda by improving conditions and offering educational opportunities for the men that were intended not only to quell further protest and maintain worker retention for its purposes—including the immediate task of procuring timber—but also to garner goodwill through benevolence. By mid-war, the Ministry of Supply's substandard treatment of the unit threatened the Dominions Office's focus. The mistreatment of the NOFU was thrown into sharp relief by comparing it to what was offered military units from other dominions, most especially those from Australia, Canada, and New Zealand. As this differential treatment suggested, the men of Newfoundland were on a different footing in the UK—as were those of British Honduras—in large part due to their home country's difficulties prior to the war, and to the responses to these difficulties, which had been used in Newfoundland to rationalize the suspension of responsible government.

Certain prevailing views of the men—as primarily labour resources and unevolved colonials with a penchant for self-produced entertainment—were factored into their treatment by those responsible for their well-being. These particular Newfoundland stereotypes—partial truths about a few that are generalized to all—are long-standing and well-documented. They had been enshrined in the *Amulree Report,* which had outlined the basis for the suspension of responsible government in 1934 (Bishop Stirling, 2001). They were evident in the attitudes of some of the NCG commissioners, as well as those of some UK civil servants, who could point not only to the rationale for the establishment of the NCG but also to the behaviour of some of the NOFU men themselves to reinforce their beliefs. Even St. John Chadwick of the Dominions Office, a supporter of and advocate for the unit, observed that "the men generally are like a bunch of overgrown children, and like them quite ready to grouse one moment and to laugh the next" (St. John Chadwick, 1942c, p. 4). His comments echoed those of the *Amulree Report*, in which the people of the country were cast as exhibiting "a childlike simplicity" that made them easily exploitable by politicians (NRC *Report*, 1933, p. 77).[2]

As a result of a series of camp and workplace improvements, from 1943 onward the NOFU's strength stabilized, productivity increased, opportunities opened for meaningful after-work community participation, and a greater sense of unit identity and purpose emerged, most especially through formal educational activities, the formation of the 3rd Inverness (Newfoundland) Battalion of the Home Guard, and the Newfoundland Overseas Foresters Association. These developments built self-confidence and solidarity among members, providing a basis of strength upon their return to Newfoundland at a time when change was afoot and their service and rights would not be acknowledged.

On their return, knowledge gain that had occurred in the UK was offset by the cultural loss and changes they found at home. During their time overseas, the home country had been transformed by Canadian and American armed forces, whose presence there drove the wartime economy and introduced enticing new ways, thereby accelerating cultural transformation on many fronts. The impact of these changes was not evenly distributed among the men of the unit. Language change and loss, for example, threatened not only local dialects of English, it also threatened Indigenous languages in Labrador and on the Island, as well as the prevalence of French in francophone communities. The ex-NOFU

2 In 1946, on behalf of the Dominions Office, St. John Chadwick and Edgar Jones of the Treasury Department prepared an advisory financial report for the deliberations of the Newfoundland National Convention. In 1967, St. John Chadwick published *Newfoundland: From Island to Province*, a constitutional history to Confederation.

members from Mi'kmaw communities on the Island returned to a way of life that was not only misunderstood but was also quickly and actively being threatened by timber harvesting that ravaged the land, destroyed natural habitats, and affected lifeways. The Innu and Inuit of Labrador faced similar threats as large-scale tree harvesting there increased after the war and Confederation.[3]

The struggle for recognition and value is a persistent theme in the story of the NOFU. While typically discussed in terms of the postwar period during which the executive of the NOFA fought for veterans' benefits through recognition of ex-NOFU members' service, the unit and its leadership had been forced to address issues of rights and recognition—and identity—from the NOFU's formation and throughout its time overseas, as well as at home in the postwar period and for the remainder of the twentieth century. Similar issues to those the NOFU men experienced had been evident in early twentieth-century woods industries, in which woods workers were persistently undervalued; they occupied the lowest rung on the industry ladder and their needs were often unheard and unaddressed. Woods workers were generally vulnerable to manipulation of their labour and to indifference toward their well-being—extensions of these pre-existing circumstances were suffered by the NOFU throughout and after its own wartime work.

Occupational Nomenclature: Another Quest for Legitimacy

During its time in the UK and afterwards, a shift in occupational nomenclature became one way in which the image of the men of the NOFU and the value of their work were adjusted. This change reflected a transition of identity—from one that was nestled in the powerlessness and impoverishment of the sort experienced by woods workers in the 1930s, to one at least partially dignified by overseas war work and the struggle for rights and recognition that began there. For the ex-NOFU men at home after the war, there was a series of identity shifts that occurred simultaneously with the transition of

3 The harvesting around Lake Melville to supply the linerboard mill at Stephenville is a particularly devastating example. See Kelly (2021a).

their homeland—from country to province—and with the related changes for its people, from subjects to citizens, and from one nationality to another.

In his biography of the first premier of Newfoundland and Labrador, Richard Gwyn recounts a story from the International Woodworkers of America union drive in the 1950s, in which a logger expressed to Landon Ladd his agreement with the platform of demands for the newest province's loggers: "But you'll never get it," the logger said, adding in explanation that "we are only loggers" (Gwyn, 1968, p. 200). The resignation revealed in this comment about his own status as a logger is unsurprising, given the history of the occupation in early to mid-twentieth-century Newfoundland. These self-deprecating words became a rallying cry to challenge such deeply embedded hierarchies during the 1959 Newfoundland Loggers' Strike, but they were also very much a poignant "mixture of dignity and powerlessness" (Sider, 2003, p. 297), a reminder that the word, like the occupation and its location—in rural contexts—garnered little or no consideration or respect.

The NOFU men who left Newfoundland were called loggers. In the UK they were most often referred to as lumberjacks. By the time they were repatriated, they called themselves foresters. Although often used synonymously, these occupational names carry different meanings and connotations. An examination of each and the historical shift in usage reflects the evolving identity of the unit, both in terms of how many of its men saw themselves as woods workers, but also how they wished to be seen, recognized, and remembered.

At home in Newfoundland, the men who joined the NOFU were most commonly known as *loggers*, a name that had supplanted but not erased *lumbermen*, which had been used to describe those who worked in the many lumbermills of the country from the mid-1800s to the early 1900s. The word *logger*, meaning "one who fells timber," was first recorded in the early 1700s; its use had increased dramatically by the late 1800s (*Oxford English Dictionary*, 2023b). It was used pervasively in Newfoundland—so widely that Governor Walwyn suggested that, regardless of the unit's official name, the NOFU badge should bear the inscription "Newfoundland Loggers Unit" since logger was the term the men themselves used to describe their occupation (Walwyn, 1940).

Lumberjack, on the other hand, has its roots in Canadian and American

English, dating to the 1860s. Its meaning is broader than that of logger: one "who fells trees, cuts them into logs, or transports them to a sawmill" (Oxford University Press, n.d.). A lumberjack is typically one who is involved in any aspect—from felling timber to making or milling lumber—of the lumbering industry. Some members of the UK press noted that the NOFU workers did not like to be called lumberjacks ("Newfoundland Loggers Clearing," 1941; "Newfoundland Loggers Helping," 1941). When questioned about the proper descriptor for the men, one spokesperson for the unit, likely a superintendent or a foreman, told a newspaper reporter in the Brechin area of Angus to use "foresters, if you like, but we don't like lumberjack" ("Angus Newfoundlanders," 1940). It was an early recorded attempt to establish a separate occupational identity for the men of the unit.

Contrary to these observations, the word *lumberjack* was commonly used in Newfoundland and is ubiquitous in the locally composed songs and stories of woods workers of the first half of the twentieth century (Kelly & Forsyth, 2018). *Lumberman* or *lumberer* were also commonly used. The dislike for lumberjack expressed by the NOFU official may have been due, in some part, to its widespread use among Canadians. Newfoundland overseas volunteers in all forces, in both the First and Second World Wars, took pains to assert their identity as men from a distinct and separate dominion than that of their closest and much larger North American neighbour. Nevertheless, the name *lumberjack* stuck and was the term that most often appeared in news headlines about the NOFU throughout the war.

Of the occupational names used to describe the NOFU workers, *forester* is the oldest. With origins in French, it refers to "an officer having charge of a forest . . . [or] one who looks after the growing timber on an estate" (Oxford University Press, n.d.). The meaning is suggestive of leadership in forest use, conservation, and sustainability. During the Second World War, under the Ministry of Labour Occupational Classifications, forester was an umbrella term, "including woodsmen, tree-fellers, and tree-loaders," and was listed under reserved occupations (Ewbank, 1940a, p. 4). Its usage could therefore have been political, and part of efforts to link the NOFU workers to these occupational classifications more explicitly, so as to discourage attrition. It was also an occupational identifier that garnered more respect in the UK than the other less familiar colonial names and thus also likely contributed to efforts to improve the status and reputation of the men.

The word *forester* is now in established use to describe the men who enlisted in the forestry units of both world wars. During the First World War, its use was sporadic, with impetus likely coming from the initial request by the UK government to Prime Minister Morris in 1917 for "a battalion of Newfoundland foresters" (Morris, 1917a, p. 4). At that time, the words *logger* and *lumberjack* would have been far less familiar in the UK context. Despite the wording of the request, *forester* was not used in the call for volunteers to enlist in the First World War. Governor W. E. Davidson and Sir Joseph Outerbridge, vice-president of the Newfoundland Patriotic Association, however, did use it ("Nfld. Foresters in the Highlands," 1917), likely after its use in relation to the men became common in the UK and perhaps to differentiate the men from soldiers—a term that, while correct, associated men more readily with the combat battalions of the Newfoundland Regiment rather than its Forestry Companies.

The use of the word *forester* elevated the work of the unit. Its inclusion in the name of its organization, the Newfoundland Overseas Foresters Association, and its publication, the *Bulletin*, may have been an attempt to standardize its usage within the NOFU. In *They Also Served*, Thomas Curran referred regularly to the men as "the Foresters" or "Newfoundland Foresters." Curran would have been familiar with that term from his prewar forestry work at Salmonier. His emphasis on the "specialized service" the NOFU workers provided in the UK was in keeping with his overall efforts to ensure the unit received the credit it deserved and that the men who volunteered with it were recognized and respected. Its usage, more generally, was an assertion of a unique identity, a setting apart and claiming of skills, expertise, and wartime experience not captured in other descriptors—in other words, it indicated a different class of woods worker.[4]

4 Songwriter Amelia Curran, granddaughter of Joseph Curran and niece of Thomas Curran, maintained the distinction in her song "1949," composed to mark the 75th anniversary of Newfoundland and Labrador's entry into Confederation. Recalling the first decade as a Canadian province and the tumultuous times of the woods industry, her lyrics say that in 1959, "foresters and loggers strike." The song was recorded by St. John's-based folk group Rum Ragged. See Opera on the Avalon (2024).

While away, the men of the NOFU learned new ways of doing woods work, including conservation practices. The UK hillsides and wooded estates were being denuded for the second time in just twenty-five years. More sustainable ways of thinking about the natural environment were emerging through the work of the Forestry Commission, formed immediately after the First World War to oversee forest use (Oosthoek, 2013). Many of its earlier forest protection ideas were reasserted in 1943, in the publication of the White Paper *Post War Forest Policy,* which also guided forestry practices after the war (Mather, 1978). In the context of a rural skilled-labour shortage, many NOFU members remained in the UK and participated in forest rebuilding in the year immediately following the war. In addition to their wartime experiences, this work impacted how they understood the significance of the forests of their homeland. Conservation, afforestation, and reforestation—embryonic concepts in Newfoundland prior to the Second World War—became more pronounced thereafter. In many cases, they were advanced by those who had worked overseas (Robertson, n.d).

After the war, as the NOFA sought recognition for ex-NOFU workers, another kind of forester was emerging in Newfoundland, of the sort that bore some resemblance to the one described in the original definition of the word. Gary Saunders began writing in local newspapers about professional foresters and the evolution of forest practices in Newfoundland in the early 1960s. In a column with the subheading "First of the Foresters," he praised the impact of Jack Turner on the redirection of forest policy, management, and use in the country that had occurred in the 1930s under the NCG (Saunders, 1960b, p. 21). He also applauded the increase in the number of professional foresters since the war and Confederation—graduates with research-based forestry degrees from Canadian universities where such specialization was offered. In 1960, there were twenty-two professional foresters in the province—of which Saunders was one—and a total of 2,700 in Canada (Saunders, 1960a).

As the wartime forester fell into the postwar shadows, the professional forester emerged. Like the foresters of the UK, who were guided by the policies of the Forestry Commission, they led the management of provincial forests for conservation and sustainability from an age in which wood and war had been seemingly immutably joined and during which large-scale industrial resource extraction and its impact on Indigenous communities and forest life went unchecked.

Waged Work and Wartime Service

Nested within and complicating efforts for recognition, including through a change of occupational nomenclature, was the debate about the meaning of wartime service. The hierarchies of war service in Newfoundland had been galvanized during the First World War and heralded through the politics of public remembrance and commemoration that followed. At the top of the hierarchy were those who served in combat roles, were uniformed, and who fought and often died in theatres of war. While other forms of service are now more widely recognized and documented, the hierarchies remain.

As the case of the ex-NOFU members demonstrates, rights and benefits after the Second World War were tied to a notion of wartime service that was synonymous with military service—enlistment in a uniformed armed force. The meaning of *veteran*—a person who retired from an armed force—was tied to these interpretations, as were prevailing notions of masculinity, which aligned the pinnacle of manliness with combat service (McIvor, 2018). In the hierarchy of war service, "the civilian male's position was more fluid, qualified by age, occupation, and geographical location" (Peniston-Bird, 2018, p. 194), all of which were liabilities for the NOFU members. Many of them were older and they had worked in an undervalued occupation. In addition, they were from a colony with contested status, which became a province considered the poor cousin within Confederation, many of whose people had, by times, been vilified in official attitudes and documents. With these additional liabilities, performance of wartime duties outside the armed forces through paid work emphasized that theirs was a service of another order of meaning, if it was considered service at all.

The men who enrolled in the NOFU at the onset of the Second World War clearly saw themselves enrolling in a wartime service, albeit for many the unit was not their first preference. The NOFU was an opportunity for some to do their bit as soon as possible, while awaiting recruitment calls for other units. For this reason, the short six-month NOFU contract was ideal—it offered an interim income at a site closer to combat as men awaited the opportunity to enlist in an armed force. While perhaps a calculated and convenient move on their part, it was neither deceitful nor disloyal. But some of the early experiences of the NOFU men in the UK were a startling wake-up call: many people there saw them as war shirkers, not war servers. Likewise, the response of the NCG to the unit's wartime efforts in its re-establishment plans—excluding them from almost all aspects—indicated that the men were seen categorially as wartime workers not warriors.

In their postwar efforts for recognition, then, the NOFA endeavoured not only to redefine the value of the NOFU members' contribution; it also contested the dominant meaning of *wartime service* and, in so doing, contributed to the term's redefinition within the province and Canada. A broadened definition of *service*—dedicating ones skills, talents, and labour to an organization designed to support a country in its quest for peace, security, and victory against threat—means many more are of service than just those who are uniformed; as a result, they too are eligible for the benefits that service bestows, even when such benefits are not clearly or evenly distributed. Thus the history of Veterans Affairs Canada has been one of vetting and acknowledging, although not without difficulties, diverse forms of services as part of an expanding social charter for Canadian veterans (Veterans Affairs Canada—Canadian Forces Advisory Council, 2004).

If, as Peter Neary says, "war made service the hallmark of citizenship, and service implied both rights and rewards" (1993a, p. 90), it was not the NOFU members' experience. For them, this idea of service was far from realized, both after the war and when they became Canadian citizens. Their efforts for recognition—suggested in the title to Thomas Curran's book about the unit, *They Also Served*—were part of how they built their concept of citizenship, which included validating their service, expanding their rights, and securing their long-overdue benefits. As part of this process, they left behind the status of "subject," and learned to inhabit a new category: citizen, with its new meanings, rights, and rewards. The ex-NOFU men, of course, were not alone in this transition.[5] The expansion of notions of service and citizenship involved more than policy shifts; it required shifts in thinking among the general population about who is a citizen, and who and what count as service to a country, and how recognition of that service is apportioned—educational work that continues to this day.

Among the veterans whose service was not contested, and within the many civilian groups whose services were not initially recognized, were Indigenous persons and others whose status was precarious for other reasons. Indigenous veterans from several Allied countries have documented the disorientation of re-entry to civilian life, and the reassertion of discriminatory attitudes toward them, that they experienced after the war (Sheffield, 2018). Canadian Indigenous veterans faced difficulties accessing benefits under the umbrella of what was then called the Indian Act (Sheffield, 2018; Veterans Affairs Canada—Canadian Forces Advisory Council, 2004). Neither the Mi'kmaq of the Island of Newfoundland nor the Innu and Inuit of mainland Labrador were recognized in the Terms of Union of 1949 between Newfoundland and Canada. Colonialism continued to complicate their efforts for recognition and reward during the postwar period in their new country, including those who were ex-NOFU members.[6] Now, reconciliatory measures—for example, National Indigenous Veterans Day, which was first observed in 1994 and is now celebrated yearly in Canada on November 8—commemorate Indigenous veterans and their contributions and assert the importance of distinct national recognition of their service.

Research as Remembrance

The failure to validate and recognize encourages forgetting, and remembrance is the resulting casualty. Storytelling can be reparative, bringing forward for

---

5 For a full account of the evolution of the Veterans Charter, see Veterans Affairs Canada—Canadian Forces Advisory Council (2004).

6 See Jeddore (2015) for an account of the postwar experiences of one Mi'kmaw ex-NOFU member.

remembrance and reconsideration the efforts and experiences of those sidelined by history, challenging the exclusions of commemoration and revealing fresh insights and meanings.

More than three-quarters of a century since the formation of the NOFU and the departure of the first contingents of workers, fuller details and also more perturbing aspects of the unit's story are evident. Archival documents related to the NOFU that are publicly available in both the UK and Newfoundland and Labrador reveal the politicized nature of the organization and its work, as well as the attitudes and actions in relation to it of those under whose authority it fell. These details suggest how the unit's formation fit within the desperate times of the Great Depression, and how the experiences of the men while overseas fell under the shadow created by the *Amulree Report* and the Commission of Government. They also reveal how the ultimate fate of the unit was tied to the currents of imperial relations of the period, including those concerning the future of Newfoundland and the British Empire after the war.

Some accounts of Second World War forestry units focus on the lack of remembrance practices for these non-combat wartime units, in particular those that were marginalized by civilian status, gender, and race. For example, the NOFU is described as "the forgotten service" (T. V. Curran, 1987, p. 66), the Women's Timber Corps as "the forgotten army" (Foat, 2019; Vickers, 2011), and the British Honduras Forestry Unit as the "forgotten Belize lumberjacks" (Quigley, 2019). These units also share another unfortunate commonality: all of them experienced inequity not only in postwar remembrance, but also in postwar support. When the war ended, their members were not eligible for pensions and other benefits.

Like the story of the Newfoundland Forestry Companies of the First World War, the narrative about the NOFU reveals much about the politics of service, recognition, and remembrance and the unit's relationship to broader social hierarchies and larger political forces. When the National War Memorial to commemorate Newfoundland's First World War effort was in the planning stages, a forester was not included in the design.[7] That initial omission was addressed, and five figures were included in the imposing sculpture unveiled in downtown St. John's in 1924. They included a woman bearing torch and sword towering atop—a symbol of freedom and empire—and four men to represent sectors of Great War service. Flanking her midway down her stone perch are a uniformed sailor and soldier; front centre, side-by-side, are a forester and a mariner.

The forester, although uniformed as a member of the Newfoundland Forestry Companies, is depicted casually, axe thrown over his shoulder, uniform shirt partly unbuttoned, sleeves rolled and exposing large biceps. The pose is suggestive of his value as a labourer, and the dressed down look serendipitously allows for the mental superimposing of an image of a civilian (NOFU) woods worker, too. The pose is suggestive of a quasi-status—the woods worker is neither fully soldier nor fully civilian. As the NOFU workers of the Second World War quickly learned—and as the men of the Newfoundland Forestry Companies of the First World War certainly knew well—the tensions between these service categories were more than representative. They informed how their work was received in the UK and valued and recognized at home.[8]

As efforts to gain meaningful recognition for the NOFU unfolded in the postwar period, and with little of the history of the unit readily available to them, families of those who had gone overseas with the unit sought answers to their questions by taking pilgrimage to UK work sites and nearby villages or, for the families of the men who had remained in the UK after the war, to

7 See Kelly (2020) for an account of the design of the National War Memorial as it relates to the inclusion of the forester figure.

8 See Kelly (2020) for an account of the pejorative attitudes of some, including soldiers of the Royal Newfoundland Regiment, towards those who served in the Newfoundland Forestry Companies. For example, one soldier of the Regiment threatened to return his medals when he learned that foresters were also eligible for (and had been awarded) some of the same medals, He questioned their eligibility and courage, accusing those who enlisted in the forestry unit of doing so in order to "take good care that they did not serve in a theatre of war at all" (Janes, 1921, p. 2). As a criterion for enlistment in the Newfoundland Forestry Companies, a man had to be ineligible—by age, physical ability, or another requirement—to enlist in the Newfoundland Regiment.

The National War Memorial, St. John's, ca. 1924. (Courtesy of Archives and Special Collections, Geography Collection 05.05.018, Memorial University of Newfoundland)

Sculptor Gilbert Bayes's model for the forester and mariner on the National War Memorial, St. John's, ca. 1923.(Courtesy of The Rooms Provincial Archives Division, B 20—137)

the homeland from whence their relatives had come. Some answers can be found in family searches, but fulsome answers require investigations into the broader and considerable domains of politics, labour, migration, citizenship, and service. Considered together, the complexity of the NOFU story, of how and why the men and their country mattered, becomes clearer.

The wartime contribution of the Newfoundland Overseas Forestry Unit is now often subsumed under Canada's contribution to the Second World War.[9] Its specificity as a unit from a then-separate dominion, albeit with a precarious place in the empire given its state of suspended self-government, is thus erased, and with it much of the significance of its contribution to that war. The NOFU was a large and largely rural-based workforce representing a

9 See, for example, the Commonwealth Foresters' Memorial. In summer 2022, the Government of Scotland and the City of Glasgow and partners dedicated the Commonwealth Foresters' Memorial at Pollock Park, Glasgow. The memorial consists of trees planted in a concentric circle to resemble the growth rings of a tree trunk and commemorates the centenary of the Forestry Commission and the forestry units who served in the UK during the Second World War (G. Davidson, 2022).

small dominion whose overall wartime service numbers nevertheless exceeded twenty thousand—a remarkable contribution. This subsuming of identity is an unfortunate historic fate for a unit whose members clearly sought recognition on several levels, including as patriots who, while away from home, imagined and planned for better times in the postwar period for themselves, their families, and their country.

Their patriotism, like their work, was not rewarded. On the heels of ill treatment as workers in the early decades of the twentieth century, and general indifference to their plight while overseas in the first years of the war, the woods workers of the Second World War were shut out by the Commission of Government and then largely ignored by the Smallwood government of their new province. In the end, it was a too-long and difficult path to rights and benefits as Canadians, one laced with contradictions and eclipses as well as points of validation, that eventually garnered for them the by-then bittersweet recognition they had sought.

REFERENCES

About ourselves. (1945, March). *Newfoundland Foresters' Bulletin*, 5, 4.

Abuse of billet allowances: Foresters refused to work—got full pay. (1941, April 24). *The Courier and Advertiser* (Dundee), 2.

Acton, H. B. (1943, October 7). [Letter to General Carrington, Home Grown Timber Production Department]. Newfoundland Forestry Unit in Scotland: welfare arrangements (AVIA 22 1352). The National Archives, London, UK.

Acton, H. B. (1944, October 7). [Letter to R. Meiggs, Home Grown Timber Supply Department]. Newfoundland Forestry Unit in Scotland: welfare arrangements (AVIA 22 1352). The National Archives, London, UK.

Address to the men of Newfoundland: Lord Northcliffe visits our foresters now in Scotland. (1918, November 5). *The St. John's Daily Star*, 3. https://collections.mun.ca/digital/collection/daily_star/id/15733/rec/2

Alas! The poor native! (1917, May 16). *The St. John's Daily Star*, 1. https://collections.mun.ca/digital/collection/daily_star/id/12844/rec/2

Alexander, D. (1980). Literacy and economic development in nineteenth century Newfoundland. *Acadiensis*, *10*(1), 3–34. https://journals.lib.unb.ca/index.php/Acadiensis/article/view/11538

Ammon, C. G. (1944). *Newfoundland, the forgotten island*. Fabian. https://collections.mun.ca/digital/collection/cns/id/128010

An Act Respecting the Newfoundland Overseas Forestry Unit. Statutes of Newfoundland 1939. c. 47, pp. 250-251 (1939, November 18). https://collections.mun.ca/digital/collections/statutesnl/id/20630/rec/1

Angus Newfoundlanders are a happy family—but don't call them lumberjacks. (1940). [newspaper clipping]. Photographs and Cuttings Related to the

Newfoundland Overseas Forestry Unit Collected by Tom Hall (PD 19 1). Aberdeen City and Aberdeenshire Archives, Aberdeen, Scotland.

Annual report. (1946). Newfoundland Commission of Government Department of Natural Resources. https://collections.mun.ca/digital/collection/annuals/id/101913/rec/41

Annual report land development for 1946. (1947, June). *Newfoundland Government Bulletin, 6.* Newfoundland Government. https://collections.mun.ca/digital/collection/nlbulletin/id/2203/rec/14

Appel, M. (1945, February 17). Questions on Newfoundland answered in Commons today. *The Evening Telegram,* 7. https://collections.mun.ca/digital/collection/telegram21/id/51846/rec/254

Appointments made by commission government. (1939, December 2). *The Daily News,* 3. https://collections.mun.ca/digital/collection/dailynews/id/129633/rec/7

Armstrong, M. F. (1940a, March 23). With the Newfoundland foresters in Scotland, *The Grand Falls Advertiser,* 1.

Armstrong, M. F. (1940b, May 18). Life in a Scottish lumber camp. *The Grand Falls Advertiser,* 3.

Army makes ersatz ghi. (1941, May 25). *Sunday Dispatch* (London), 7.

Arrangements for signing on men forestry unit. (1941, May 6). *The Daily News,* 3. https://collections.mun.ca/digital/collection/dailynews/id/115810/rec/455

Ashton, J. (1986). The lumbercamp song tradition in Newfoundland. *Newfoundland [and Labrador] Studies,* 2(2), 213–231. https://journals.lib.unb.ca/index.php/NFLDS/article/view/580

Attlee, C. R. (1933, December 12). "Newfoundland Bill." UK Parliament. House of Commons. *Hansard,* Vol. 284(224). https://api.parliament.uk/historic-hansard/commons/1933/dec/12/newfoundland-bill

Attlee, C. R. (1942a, March 6). [Letter to Andrew Duncan, Minster of Supply]. Newfoundland Forestry Unit in Scotland: welfare arrangements (AVIA 22 1352). The National Archives, London, UK.

Attlee, C. R. (1942b, May 28). [Letter to Andrew Duncan, Minister of Supply]. Newfoundland Forestry Unit in Scotland: welfare arrangements (AVIA 22 1352). The National Archives, London, UK.

Attlee, C. R. (1942c, July 24). [Telegram to Governor Walwyn]. Governor's Office – Local and Miscellaneous Correspondence 1942—Forestry (GN 1 /3/ A, Box 245, File 2). The Rooms Provincial Archives, St. John's, NL, Canada.

Attlee, C. R. (1942d, August 25). [Letter to Gwilym Lloyd George, Ministry of Supply]. Newfoundland Forestry Unit in Scotland: welfare arrangements (AVIA 22 1352). The National Archives, London, UK.

Attlee, C. R. (1942e, September 18). Rt. Hon. C. R. Attlee delivers address. *The Daily News*, 3. https://collections.mun.ca/digital/collection/dailynews/id/113362/rec/2

Attlee, C. R. (1942f, December 15). [Telegram copy to Governor Walwyn: Mr. Tait to Commissioner Dunn]. Governor's Office – Local and Miscellaneous Correspondence 1942 (GN 1/3/A, Box 242, File 2). The Rooms Provincial Archives, St. John's, NL, Canada.

Awarded MBE for wartime work. (1946, July 19). *The Grantham Journal*, 3.

Back home to stay. (1984). *Decks Awash, 13*(4), 48. https://collections.mun.ca/digital/collection/cns_decks/id/5706/rec/1

Backwoods wedding reception. (1941, August 20). *Press and Journal* (Aberdeen), 2.

Baird, J. (1981). *The Newfoundland Overseas Forestry Unit* [unpublished manuscript]. Folklore and Language Archives, 81–428, Memorial University, St. John's, NL, Canada.

Baker, G. S. (2000, September 22). "Civilian War-Related Benefits Act." Canada. Parliament. House of Commons. *Edited Hansard, 120*(1010). 36th Parliament, 2nd session. House Publications: https://www.ourcommons.ca/documentviewer/en/36-2/house/sitting-120/hansard#LINK5

Baker, M. (2003). Falling into the Canadian lap: The Confederation of Newfoundland and Canada 1945–1949. *Royal Commission on Renewing and Strengthening Our Place in Canada* (33–88). https://www.gov.nl.ca/publicat/royalcomm/research/fallingintothecanadianlap.pdf

Baker, M. (2017). Warfare 1914–1918 (Newfoundland). In U. Daniel, P. Gatrell, O. Janz, H. Jones, J. Keene, A. Kramer, & B. Nasson (Eds.),

1914–1918-online: International encyclopedia of the First World War. Freie Universität Berlin. https://doi.org/10.15463/ie1418.10804/1.1

Baker, M., & Miller Pitt, J. (1996). Prominent figure from our recent past: Jack Turner. *The Newfoundland Quarterly*, *90*(2), 24. https://collections.mun.ca/digital/collection/quarterly/id/39051/rec/2

Banks, I. (2011). Control or repression: Contrasting a prisoner of war camp and a work camp from World War Two. In A. Myers & G. Moshenska (Eds.), *Archaeologies of internment* (pp. 111–128). Springer. https://doi.org/10.1007/978-1-4419-9666-4_7

Banquet to close Legion convention. (1983, June 22). *The Western Star*, 1. https://collections.mun.ca/digital/collection/westernstar/id/190515/rec/4

Barrett, B., Dicks, E., Brown, I., Murray, H. C., & Porter, H. F. (1996). *They came from over the sea: British war brides in Newfoundland*. British War Brides Association of Newfoundland and Labrador.

Barrett, R. (1940, March 30). Newfoundland Forestry Unit: Diary of trip. *The Grand Falls Advertiser*, 1.

Bigamy is Britain's war crime no. 1. (1942, August 21). *The Western Star*, 4. https://collections.mun.ca/digital/collection/westernstar/id/13441/rec/2

Bilgrami, S. (Director). (2004). *Tree fellers* [Film]. Asylum Pictures. https://vimeo.com/364780261

Bill C-41: An Act to Amend the Statute Law in Relation to Veterans' Benefits (1999). 36th Parliament, 2nd session. https://www.parl.ca/Content/Bills/362/Government/C-41/C-41_3/90118bE.html

Bill C-180: Appropriation Act, No. 7 (1967). 27th Parliament, 2nd session. Canadian Parliamentary Historical Resources: https://parl.canadiana.ca/view/oop.bills_HOC_2702_2/298

Bird, C. W., & Davies, J. B. (1919). *The Canadian Forestry Corps: Its inception, development and achievements*. Board of Trade and H. M. Stationery Office. https://dn790008.ca.archive.org/0/items/canadianforestry00birduoft/canadianforestry00birduoft.pdf

Bishop Stirling, T. (2001). The Amulree report: An introduction. In G. Fizzard (Ed.), *Amulree's legacy: Truth, lies and consequences symposium, March 2000* (pp. 13–26.). Newfoundland Historical Society. https://collections.mun.ca/digital/collection/cns2/id/21752/rec/4

Blackmore, J. (1946, September 20). Interesting letter from Newfoundland [Letter to the editor]. *West Lothian Courier*, 2.

Blake, R. B., & Baker, M. (2019). *Where once they stood: Newfoundland's rocky road towards Confederation*. University of Regina Press.

Boyko, J. (2021, August 3). British subject status. In *The Canadian encyclopedia*. https://www.thecanadianencyclopedia.ca/en/article/british-subject-status

Boys of Newfoundland Overseas Forestry Unit. (1940, August 28). [Letter to the editor]. *The Daily News*, 4. https://collections.mun.ca/digital/collection/dailynews/id/121720/rec/270

The boys overseas prefer Wings. (1940, June 25). *The Daily News*, 1. https://collections.mun.ca/digital/collection/dailynews/id/121123/rec/

Brass, W. (1944, April 27). [Report: Letter Enclosure from P. Emrys-Evans to C. U. Peat]. Newfoundland Forestry Unit in Scotland: welfare arrangements (AVIA 22 1352). The National Archives, London, UK.

Brewis, W. (1943, February 1). [Letter to F. M. Peters]. Photographs and Cuttings Related to the Newfoundland Overseas Forestry Unit Collected by Tom Hall (PD 19 1). Aberdeen City and Aberdeenshire Archives, Aberdeen, Scotland.

Britain's need for skilled tree fellers. (1940, April 20). *The Evening Telegraph* (Dundee), 6.

British parliamentarians visit west coast. (1943, July 17). *The Western Star*, 1. https://collections.mun.ca/digital/collection/westernstar/id/14099/rec/13

Bromley, A. (1940, March 10). *Visit to Newfoundland loggers, March 10th, 1940* [unpublished report enclosed in a letter from R. Ewbank to R. Robinson, 1940, March 21]. Dominions Office, Original Correspondence, 1943–46 (DO 35, Reel B 5020 /744, File N.264/30–50). The Rooms Provincial Archives, St. John's, NL, Canada.

Brooks, A. J. (1960, April 1). [Letter to Premier Smallwood]. J. R. Smallwood Papers, Collection 075, Box 191, File 3.10.031). Archives and Special Collections, Memorial University of Newfoundland, St. John's, NL, Canada.

Browne, R. (1945, July). Letter from the Association to the Commission of Government, Newfoundland. *Newfoundland Foresters' Bulletin*, 9, 3.

Bruce, R. (1945, June 6). What I think of you. *The Daily News*, 7. https://collections.mun.ca/digital/collection/dailynews/id/100257/rec/1

Bruyea, S., & Smol, R. (2016, December 15). Calling out the great veteran pretender. *Esprit de corps: Canadian Military Magazine*. https://www.espritdecorps.ca/perspectives-1/calling-out-the-great-veteran-pretender

Burns, E. L. M. (August 4, 1950). [Letter to Isaac Mercer]. J. R. Smallwood Papers, (Collection 075, Box 144, File 3.03.012). Archives and Special Collections, Memorial University of Newfoundland, St. John's, NL, Canada.

Busch, B. C. (1984, Fall). The Newfoundland sealers' strike of 1902. *Labour/Le Travailleur 14*, 73–101. https://id.erudit.org/iderudit/llt14art03

Butler, E. (1941, June 27). [Letter to Tom S. Hall, Newfoundland Forestry Unit]. Photographs and Cuttings Related to the Newfoundland Overseas Forestry Unit Collected by Tom Hall (PD 19(1)). Aberdeen City and Aberdeenshire Archives, Aberdeen, Scotland.

Butler, T. (1940, May 21). With the foresters [Letter to the editor]. *The Evening Telegram*, 5. https://collections.mun.ca/digital/collection/telegram21/id/30300/rec/16

Butler, T. (1943). *Blueprint for Newfoundland: An industrial prospectus for Newfoundland's development*. Author.

Butler, T. (1945, March). Letter to the editor. *Newfoundland Foresters' Bulletin*, 5, 2.

Butler, V. L. (1941, June 28). A letter from one of "ours" overseas. *The Newfoundland Weekly*, 8. https://collections.mun.ca/digital/collection/nfldweekly/id/2275/rec/4

Cadigan, S. T. (2006). Regional politics are class politics: A Newfoundland and Labrador perspective on regions. *Acadiensis*, 35(2), 163–168. https://www.jstor.org/stable/30303300

Cadigan, S. T. (2009). *Newfoundland and Labrador: A history*. University of Toronto Press.

Cadigan, S. T. (2016). Commemoration, cult of the fallen (Newfoundland). In U. Daniel, P. Gatrell, O. Janz, H. Jones, J. Keene, A. Kramer, and B. Nasson (Eds.), *1914–1918-online: International encyclopedia of the First World War*. Freie Universität Berlin. https://doi.org/10.15463/ie1418.10955

Cameron, I. (2008). *Plants and roots: A social history of Ballater before the Beatles and Dr. Beeching*. Pica Design.

Cameron, I. (2010). *Albert Road . . . the place of my birth*. Pica Design.

Cameron, I. (2018). The Deeside forestry scene in wartime Scotland. In U. A. Kelly & M. C. Forsyth (Eds.), *The music of our burnished axes: Songs and stories of the woods workers of Newfoundland and Labrador* (pp. 404–407). ISER Books / Memorial University Press.

Camp meeting at Deisher. (1945, March 15). *The Daily News*, 9. https://collections.mun.ca/digital/collection/dailynews/id/98016/rec/65

Canada is seeking loggers in Nfld. (1943, May 15). *The Evening Telegram*, 7. https://collections.mun.ca/digital/collection/telegram21/id/41720/rec/39

Canada. Parliament. House of Commons. Standing Committee on Veterans Affairs. (1960). *Minutes of proceedings and evidence* (No. 1, 24th Parliament, 3rd session). Canadian Parliamentary Historical Resources. https://parl.canadiana.ca/view/oop.com_HOC_2403_15_1/103

Canada. Parliament. Senate. Subcommittee on Veterans Affairs. (1991, January). *It's almost too late: Report of the subcommittee on Veterans Affairs of the standing committee on social affairs, science and technology*. Canadian Parliamentary Historical Resources: https://parl.canadiana.ca/view/oop.com_SOC_3402_15_2/8

Canadian beaus top marriage poll—3700 chose Scots girls. (1945, February 17). *Daily Record*, 5.

Canadian Legion conference. (1960, June 29). *The Daily News*, 9. https://collections.mun.ca/digital/collection/dailynews/id/30588/rec/13

Carter, C. (1962, February 14). "Civilian War Pensions." Canada. Parliament. House of Commons. *Debates Hansard*, 1(863–870). 24th Parliament, 5th session. Canadian Parliamentary Historical Resources: https://parl.canadiana.ca/view/oop.debates_HOC2405_01/865

Carter, H. (1943). *Victory songs and other verse*. Long Brothers. https://collections.mun.ca/digital/collection/cns/id/89465/rec/1

Carter, K. J. (1943, June 19). Appeal for binoculars. *The Western Star*, 7. https://collections.mun.ca/digital/collection/westernstar/id/14053/rec/13

Casey, G. J., & Hanrahan, M. C. (1994). Roses and thistles: Second World War brides in Newfoundland. *Newfoundland [and Labrador] Studies, 10*(2), 240–249.

Charges foresters are excluded from charter. (1963, August 16). *The Daily News*, 3. https://collections.mun.ca/digital/collection/dailynews/id/45507/rec/35

Churchill, G. (1962, December 17). "Newfoundland Overseas Forestry Unit: Inclusion under Veterans Charter." Canada. Parliament. House of Commons. *Debates Hansard*, 3(2708–2709). 25th Parliament, 1st session. Canadian Parliamentary Historical Resources: https://n2t.net/ark:/69429/m0k06ww7b81x

Claude Fraser: Obituary. (1945). *The Adelphian*, 45, 151–153. https://collections.mun.ca/digital/collection/adelphian/id/6069/rec/18

Cochrane, J. A. (1945, March 20). "When the boys come home" [Interview]. *The Daily News*, 5. https://collections.mun.ca/digital/collection/dailynews/id/98059/rec/1

Colbourne, R. T. (1961, April 15). The Legion corner: Letter to F. G. Matthews. *The Western Star*, 6. https://collections.mun.ca/digital/collection/westernstar/id/62230/rec/2

Collins, P. W. (2014). Not a strike but a civil war: The International Woodworkers of America (IWA) loggers' strike and the Badger riot 10 March 1959. *Heritage Newfoundland and Labrador*. https://heritagenl.ca/wp-content/uploads/2020/05/21-The-Badger-Riot.pdf

Collins, W. (1963, October 5). Questions foresters claim [Letter to the editor]. *The Daily News*, 4. https://collections.mun.ca/digital/collection/dailynews/id/46194/rec/2

Colonial foresters. (1943, May 8). *The Scotsman*, 4.

Comforts from W.P.A. received overseas. (1944, February 22). *The Daily News*, 5. https://collections.mun.ca/digital/collection/dailynews/id/103003/rec/306

Concert bus. (1940, April 30). *Edinburgh Evening News*, 3.

Crews, N. (1965, September 9). The foresters' case [Letter to the editor]. *The Evening Telegram*, 6–7. https://collections.mun.ca/digital/collection/telegram21/id/193684/rec/1

Cribb, M. (1946, July 21). It's a grand life, says Myra [Letter to the editor]. *The Sunday Post*, 4.

Crocker, M. (1996). Untitled. In B. Barrett et al (Eds.), *They came from over the sea: British war brides in Newfoundland* (pp. 22–23). British War Brides Association of Newfoundland and Labrador.

Curran, J. (1940, June 7). *Supplement to monthly report of March 31, 1940, on Newfoundland Overseas Forestry Unit*. Commission of Government, Natural Resources, General Administration 1939–1942 (GN 38.2, S2–1-17, File 35). The Rooms Provincial Archives, St. John's, NL, Canada.

Curran, T. V. (1987). *They also served: The Newfoundland Overseas Forestry Unit 1939–1946*. Jesperson.

Curtis, B. (1990). Some recent work on the history of literacy in Canada. *History of Education Quarterly, 30*(4), 613–624. https://doi.org/10.2307/368949

The *Daily News* diary of the war in 1942—Events of the year at home and abroad. (1942, December 31). *The Daily News*, 66. https://collections.mun.ca/digital/collection/dailynews/id/114519/rec/1

Davidson, G. (2022, October 3). Unsung forestry heroes from the Commonwealth recognised in living memorial in Glasgow. *The Scottish Farmer*. https://www.thescottishfarmer.co.uk/news/22972020.unsung-forestry-heroes-commonwealth-recognised-living-memorial-glasgow/

Davies, D. J. (1945, October 11). [Letter to Secretary for Home Affairs]. Imperial War Graves Commission (IWGC), Newfoundland Forestry Unit, 11 October 1945–13, December 1952 (GN 2.5.464.12). The Rooms Provincial Archives, St. John's, NL, Canada.

Day, D. (1965, July 7). Waiting 20 years . . . foresters still hopeful. *The Evening Telegram*, 4. https://collections.mun.ca/digital/collection/telegram21/id/193655/rec/1

Deeside lumbermen on the air. (1941, March 4). *Press and Journal* (Aberdeen), 3.

Disorderly lumbermen at Kirriemuir. (1941, October 16). *Kirriemuir Free Press and Angus Advertiser*, 1.

Dominions Office. (1940, April 24). *Minutes of meeting at the Dominions Office regarding the situation among the [NF] loggers. Dominions Office*, Original Correspondence, 1943–46 (DO 35, Reel B 5021 /745, File N.264/52–213). The Rooms Provincial Archives, St. John's, NL, Canada.

Dooley, J. P. (1978). *Tapes, manuscript notes and transcripts from collecting project sponsored by St. John's Folk Arts Council, Summer, 1977* [Unpublished manuscript and tapes]. Folklore and Language Archives, A78–167, C3676-C3682, Memorial University, St. John's, NL, Canada.

Doyle, G. S. (1940). *The old-time songs and poetry of Newfoundland* (2nd ed.). Gerald S. Doyle. https://collections.mun.ca/digital/collection/cns/id/69795/rec/1

Duke of Devonshire. (1940, April 23). [Minutes on visit to NOFU camps]. Dominions Office, Original Correspondence, 1943–46 (DO 35, Reel B 5020 /744, File N.264/30–50). The Rooms Provincial Archives, St. John's, NL, Canada.

Duncan, A. R. (1942). [Memorandum to Secretary of State for Dominion Affairs]. Newfoundland Forestry Unit in Scotland: welfare arrangements (AVIA 22 1352). The National Archives, London, UK.

Dunn, P. D. H. (1941a, September 25). *Memo submitted for consideration of Commission of Government by Commissioner of Natural Resources (Workers Compensation*. Commission of Government, Natural Resources, General Administration 1940–1941 (GN 38.2, S2–1-20, File 7). The Rooms Provincial Archives, St. John's, NL, Canada.

Dunn, P. D. H. (1941b, December 31). *Memo submitted for consideration of Commission of Government by Commissioner of Natural Resources (Essential Work Order)*. Commission of Government, Natural Resources, General Administration 1940–1941 (GN 38.2, S2–1-20, File 7). The Rooms Provincial Archives, St. John's, NL, Canada.

Dunn, P. D. H. (1942, April 24). *Memo regarding liability of NCG for repatriation of NOFU members*. Commission of Government, Natural Resources, Forestry Division (GN 38.2, S2–3-3, File 1). The Rooms Provincial Archives, St. John's, NL, Canada.

Dunn, P. D. H. (1945a, May 5). Address by Hon. P. D. H. Dunn, O.B.E. *The Western Star*, 10, 12–13. https://collections.mun.ca/digital/collection/westernstar/id/15661/rec/226

Dunn, P. D. H. (1945b, July). Reply from the Commissioner of Natural Resources. *Newfoundland Foresters' Bulletin*, 9, 3.

Dupuis, S. (2021). Francophones of Newfoundland and Labrador. In *The Canadian encyclopedia*. https://www.thecanadianencyclopedia.ca/en/article/francophones-de-terre-neuve-et-labrador

Eastment, D. J. (1982). *The policies and position of the British Council from the outbreak of war to 1950* [Doctoral dissertation, University of Leeds]. British Library EThOS. https://etheses.whiterose.ac.uk/id/eprint/640/

Eden, A. (1939, November 9). [Telegram to Governor H. E. Walwyn]. Governor's Office – Local and Miscellaneous Correspondence 1936—Forestry Recruitment (GN 1/3/A, Box 187, File 118). The Rooms Provincial Archives, St. John's, NL, Canada.

Eden, A. (1940, April 25). [Telegram to Governor H. E. Walwyn]. Governor's Office – Local and Miscellaneous Correspondence 1936—Forestry Recruitment (GN 1/3/A, Box 187, File 118). The Rooms Provincial Archives, St. John's, NL, Canada.

Editorial. (1945a, January). *Newfoundland Foresters' Bulletin*, *3*, 1. Photographs and Cuttings Related to the Newfoundland Overseas Forestry Unit Collected by Tom Hall (PD 19). Aberdeen City and Aberdeenshire Archives, Aberdeen, Scotland.

Editorial. (1945b, July). *Newfoundland Foresters' Bulletin*, *9*, 1.

Editorial. (1946, April). *Newfoundland Foresters' Bulletin*, *18*, 1. http://nofu.ca/submitted_items/skenway_bulletin_p1.html

Emerson, L. E. (1942, December 9). Text of honourable L. E. Emerson's broadcast address given last night. *The Evening Telegram*, 3–4. https://collections.mun.ca/digital/collection/telegram21/id/41240/rec/1

Emrys-Evans, P. V. (1942a, May 21). [Letter to Clement Attlee]. Dominions Office, Original Correspondence, 1943–46 (DO 35, Reel B 2511 /1155, File N.664/4–65). The Rooms Provincial Archives, St. John's, NL, Canada.

Emrys-Evans, P. V (1942b, May 26). *Newfoundland Forestry Unit* (Enclosure: Superintendents' Resolutions). Dominions Office, Original Correspondence, 1943–46 (DO 35, Reel B 2511 /1155, File N.664/4–65). The Rooms Provincial Archives, St. John's, NL, Canada.

Emrys-Evans, P. V. (1942c, June 23). [Letter to Charles Peat, Ministry of Supply]. Newfoundland Forestry Unit in Scotland: welfare arrangements (AVIA 22 1352). The National Archives, London, UK.

Emrys-Evans, P. V. (1942d, September 2). *Report on visit in August 1942*. Dominions Office, Original Correspondence, 1943–46 (DO 35, Reel B 2511 /1155, File N.664/4–65). The Rooms Provincial Archives, St. John's, NL, Canada.

Emrys-Evans, P. V. (1942e, October 14). [Letter to Charles Peat, Parliamentary Secretary to Minister of Supply]. Newfoundland Forestry Unit in Scotland: welfare arrangements (AVIA 22 1352). The National Archives, London, UK.

Emrys-Evans, P. V. (1942f, November 26). [Letter to J. Turner, Officer-in-Charge, NOFU]. Newfoundland Forestry Unit in Scotland: welfare arrangements (AVIA 22 1352). The National Archives, London, UK.

Entertainment for lumberjacks. (1940, October 19). *Perthshire Advertiser*, 11.

Ewbank, R. (1940a, April 5). [Letter to Forestry Commission Chair R. Robinson]. Dominions Office, Original Correspondence, 1943–46 (DO 35, Reel B 5021 /745, File N.264/52–213). The Rooms Provincial Archives, St. John's, NL, Canada.

Ewbank, R. (1940b, June 11). [Minutes]. Dominions Office, Original Correspondence, 1943–46 (DO 35, Reel B 5021 /745, File N.264/52–213). The Rooms Provincial Archives, St. John's, NL, Canada.

Ewbank, R. (1940c, August 20). [Letter to J. R. Gorvin, Commissioner of Natural Resources]. Commission of Government, Natural Resources—General Administration 1940 (GN 38.2, S2–1-17). The Rooms Provincial Archives, St. John's, NL, Canada.

Ex-foresters association holds meeting. (1946, October 11). *The Western Star*, 1. https://collections.mun.ca/digital/collection/westernstar/id/17152/rec/6

Expresses thanks to forestry unit. (1946, July 22). *The Evening Telegram*, 3. https://collections.mun.ca/digital/collection/telegram21/id/54877/rec/1

Far-from-Home Guard stands down: Newfoundland at war. (1945, January 8). *The Daily News*, 7. https://collections.mun.ca/digital/collection/dailynews/id/97590/rec/240

Fighting for the Red Cross: Newfoundlanders change axes for boxing gloves. (1941). [unpublished manuscript]. Photographs and Cuttings Related to the Newfoundland Overseas Forestry Unit Collected by Tom Hall

(PD 19 1). Aberdeen City and Aberdeenshire Archives, Aberdeen, Scotland.

Fighting sons of Newfoundland carry on the tradition: "Better than the best." (1940, December 31). *The Daily News*, 3. https://collections.mun.ca/digital/collection/dailynews/id/123004/rec/1

Finale. (n.d.). [Lyrics and stage directions]. Photographs and Cuttings Related to the Newfoundland Overseas Forestry Unit Collected by Tom Hall (PD 19 1). Aberdeen City and Aberdeenshire Archives, Aberdeen, Scotland.

Finch, C. S. (1940, March 1). Hospitality wanted for lumberjacks [Letter to the editor]. *The Hawick News*, 4.

Fisher, A. (2009). *A story of engagement: The British Council 1934–2009*. Counterpoint. https://universityofleeds.github.io/philtaylorpapers/pmt/exhibits/3082/BritishCouncil.pdf

Fitzgerald, T. (1942a, May 27). [Letter to Sir William Palmer, Ministry of Supply]. Newfoundland Forestry Unit in Scotland: welfare arrangements (AVIA 22 1352). The National Archives, London, UK.

Fitzgerald, T. (1942b, July 2). [Letter to J.L. Sears, Ministry of Supply]. Newfoundland Forestry Unit in Scotland: welfare arrangements (AVIA 22 1352). The National Archives, London, UK.

Flynn, E. J. (1946, April). Letter to the editor. *Newfoundland Foresters' Bulletin, 18*, 1–2. http://nofu.ca/submitted_items/skenway_bulletin_p1.html

Foat, J. (2019). *Lumberjills: Britain's forgotten army*. The History Press.

Forbes, A. (2015). *The British Empire and the war effort: A comparative study of the experiences of the British Honduran Forestry Unit and the Newfoundland Overseas Forestry Unit* [Honours dissertation, University of the Highlands and Islands.] Doczz.net. https://tinyurl.com/5yjyk9y2

Ford, A. (1985). *Telling the truth. British Honduras Forestry Unit in Scotland 1941–1944*. Karia Press.

Foresters association to function in Newfoundland. (1946, April 26). *The Grand Falls Advertiser*, 1.

Foresters association wants members get volunteer medals. (1982, August 16). *The Western Star*, 5. https://collections.mun.ca/digital/collection/westernstar/id/185377/rec/1

Foresters face the future [Editorial]. (1945, October 26). *The Western Star*, 2. https://collections.mun.ca/digital/collection/westernstar/id/16099/rec/205

Foresters not included. (1945, July 31). *Observer's Weekly*, 4–5. https://collections.mun.ca/digital/collection/observersweek/id/12041/rec/2

Foresters not returning. (1941, March 5). *The Evening Telegram*, 6. https://collections.mun.ca/digital/collection/telegram21/id/35510/rec/87

Foresters petitioning with 'just, fair request.' (1966, August 19). *The Evening Telegram*, 7. https://collections.mun.ca/digital/collection/telegram21/id/197737/rec/1

The foresters return: Editorial comment. (1946, April 20). *The Grand Falls Advertiser*, 2.

Foresters seek benefits accorded other war veterans. (1963, April 5). *The Evening Telegram*, 4. https://collections.mun.ca/digital/collection/telegram21/id/173229/rec/5

Foresters still pressing to have 'injustice' righted. (1964, October 30). *The Evening Telegram*, 4. https://collections.mun.ca/digital/collection/telegram21/id/183289/rec/1

Foresters surprise paymaster. (1943, April 1). *Press and Journal* (Aberdeen), 4.

Foresters waiting for war-time benefits. (2003, November 11). CBC News. https://www.cbc.ca/news/canada/newfoundland-labrador/foresters-waiting-for-war-time-benefits-1.389241

Forestry camps: Question of sanitation. (1941, May 1). *The Scotsman*, 3.

Forestry troops will return to Canada. (1943, November 6). *The Western Star*, 9. https://collections.mun.ca/digital/collection/westernstar/id/14342/rec/167

Forestry unit and no. families taken off relief. (1940, March 23). *The Daily News*, 6. https://collections.mun.ca/digital/collection/dailynews/id/120180/rec/72

Forestry unit purchases of war savings. (1943, September 7). *The Daily News*, 4. https://collections.mun.ca/digital/collection/dailynews/id/109974/rec/269

Forestry unit sends donation. (1943, October 8). *The Evening Telegram*, 3. https://collections.mun.ca/digital/collection/telegram21/id/44462/rec/1

Forestry unit will be disbanded March 31. (1946, March 8). *The Western Star*, 1. https://collections.mun.ca/digital/collection/westernstar/id/16476/rec/37

Fraser, A. M. (2010). *History of the participation by Newfoundland in World War II* [unpublished manuscript, P. Neary & M. Baker, Eds.]. Centre for Newfoundland Studies, Memorial University, St. John's, NL. https://collections.mun.ca/digital/collection/cns_publish/id/208

Fraser, C. (1940, August 17). [Letter to Commissioner J. H. Gorvin: Report on the formation, organization, enlistment, and transportation of the Newfoundland Forestry Unit]. Secretary of the Commission of Government—Natural Resources (GN38, Box S2–1-18, File 6). The Rooms Provincial Archives, St. John's, NL, Canada.

Further appreciation: Tribute to the foresters. (1919, March 18). *The Evening Telegram*, 9. https://collections.mun.ca/digital/collection/telegram20/id/680/rec/1

Gabriel, N. (1999). I called the wilderness mother wild. In *Not too long ago . . . : Seniors tell their stories* (pp. 94–95). The Seniors Resource Centre, St. John's, NL.

Gallacher, W. (1945, February 13). "Newfoundland." UK Parliament. House of Commons. *Debates Hansard*, Vol. 408(14). https://api.parliament.uk/historic-hansard/commons/1945/feb/13/newfoundland#S5CV0408P0_19450213_HOC_87

Gaunt, J. R. (1940, May 2). [Letter to D. J. Davies re badges, Newfoundland Forestry Unit]. Dominions Office, Original Correspondence, 1943–46 (DO 35, Reel B 5021 /745, File N.264/52–213). The Rooms Provincial Archives, St. John's, NL, Canada.

Gladney, L. (1999). *Lumberjack Larry: Memoirs of Larry Gladney*. Natural Media.

Goobie, J. (n.d.). *From Josh Goobie*. Newfoundland Overseas Forestry Unit 1939–1945. http://nofu.ca/submitted_items/josh_goobie.html

Gorvin, J. H. (1939, November 20). Nfld. Overseas Forestry Unit to be formed: Mr. Gorvin's address. *The Daily News*, 3, 6. https://collections.mun.ca/digital/collection/dailynews/id/129504/rec/2

Gorvin, J. H. (1940a, January 23). [Telegram to the Secretary of State for Dominion Affairs]. Governor's Office – Local and Miscellaneous

Correspondence 1941—Forestry Recruitment (GN 1/3/A, Box 187, File 118). The Rooms Provincial Archives, St. John's, NL, Canada.

Gorvin, J. H. (1940b, July 4). [Memo to Newfoundland Commissioners]. Governor's Office – Local and Miscellaneous Correspondence 1941—Forestry Recruitment (GN 1/3/A, Box 187, File 118). The Rooms Provincial Archives, St. John's, NL, Canada.

Gorvin, J. H. (1940c, November 28). [Memo re analysis of reasons given by NOFU men for leaving unit]. Commission of Government, Natural Resources—General Administration (GN 38.2, S2–1-17, File 25). The Rooms Provincial Archives, St. John's, NL, Canada.

Goss, M. (1980). *The role of tradition in the personal life of Josephine Goss, RN* [unpublished manuscript]. Folklore and Language Archives, 81–186, Memorial University of Newfoundland, St. John's, NL, Canada.

Great War Veterans' Association. (1940). *Report of the Dominion executive.* https://collections.mun.ca/digital/collection/cns/id/71145/rec/8

Green, L. (2023). *Where once they logged.* DRC Publishing.

Green, L. (2025). *Shattered dreams: Stories of loggers buried overseas.* Amazon Kindle Direct.

Greene, A. (1962, October 18). Fairness for foresters. [Letter to the editor]. *The Evening Telegram*, 6, 13. https://collections.mun.ca/digital/collection/telegram21/id/167503/rec/66

Greene, A. (1963, October 12). Unjust to foresters. [Letter to the editor]. *The Daily News*, 4. https://collections.mun.ca/digital/collection/dailynews/id/43046/rec/26

Greene, A. (1965, September 28). Foresters and rights. [Letter to the editor]. *The Evening Telegram*, 6. https://collections.mun.ca/digital/collection/telegram21/id/194155/rec/1

Gregg, M. F. (1950a, July 7). [Letter to Premier Smallwood]. J. R. Smallwood Papers, Collection 075, Box 144, File 3.03.012). Archives and Special Collections, Memorial University of Newfoundland, St. John's, NL, Canada.

Gregg, M. F. (1950b, July 20). [Letter to Premier Smallwood]. J. R. Smallwood Papers, (Collection 075, Box 144, File 3.03.012). Archives and Special Collections, Memorial University of Newfoundland, St. John's, Canada.

Gregg, M. F. (1950c, July 25). [Letter to Premier Smallwood]. J. R. Smallwood Papers, (Collection 075, Box 144, File 3.03.012). Archives and Special Collections, Memorial University of Newfoundland, St. John's, Canada.

Griffin, B. (2024). Perceptions of crisis in the history of masculinity: Power and change in modern Britain. In M. Houlbrook, K. Jones & B. Mechen (Eds.), *Men and masculinities in modern Britain: A history for the present* (pp. 158–177). Manchester University Press. https://doi.org/10.7765/9781526174703.00013

Gullace, N. F. (1997, April). White feathers and wounded men: Female patriotism and the memory of the Great War. *Journal of British Studies, 36*(2), 178–206. http://www.jstor.org/stable/176011

Gunston, D. (1943, December 16). "Newfoundland." UK Parliament. House of Commons. *Debates Hansard*, Vol. 395(1743–87). https://api.parliament.uk/historic-hansard/commons/1943/dec/16/newfoundland#column_1758

Gushue, R. (1956, September). The Newfoundland Woods Labour Board. *Relations industrielles/Industrial Relations, 11*(4), 268–279. https://doi.org/10.7202/1022611ar

Gwyn, R. (1968). *Smallwood: The unlikely revolutionary*. McCLelland & Stewart.

Hall, T. (1941). *The two Kildrummies* [unpublished manuscript]. Photographs and Cuttings Related to the Newfoundland Overseas Forestry Unit Collected by Tom Hall (PD 19 2). Aberdeen City and Aberdeenshire Archives, Aberdeen, Scotland.

Hanley, A. (2022). Migration, racism and sexual health in postwar Britain. *History Workshop Journal, 94*, 202–222. https://doi.org/10.1093/hwj/dbac018

Hanrahan, M. (2013). Indigenizing the academy: An institutional case study of one university. *The Morning Watch, 40*(3–4), 104–115. www.mun.ca/educ/faculty/mwatch/vol40/winter2013/caseOfOne.pdf

Harding, R. J. (2001). Glorious tragedy: Newfoundland's cultural memory of the attack at Beaumont Hamel, 1916–1925. *Newfoundland and Labrador Studies, 21*(1), 3–40. https://journals.lib.unb.ca/index.php/NFLDS/ article/view/5884

Hart, A. E. (1940, September 9). [Memo to Forestry Commissioner re welfare activities]. Dominions Office, Original Correspondence,

1943–46 (DO 35, Reel B 5021 /745, File N.264/52–213). The Rooms Provincial Archives, St. John's, NL, Canada.

Hart, A. E. (1941a, August 12). [Memo to Deputy Director of Timber Supply on behalf of Officer-in-Charge Turner]. Dominions Office, Original Correspondence, 1943–46 (DO 35, Reel B 5021 /745, File N.264/52–213). The Rooms Provincial Archives, St. John's, NL, Canada.

Hart, A. E. (1941b, September 25). [Memorandum to J. Turner, J.M. Curran, and Superintendent—Kershopefoot: Rosebery Camp]. Maurice Job Taylor Fonds (MG 390.49, Box 2). The Rooms Provincial Archives, St. John's, NL, Canada.

Hart, A. E. (1941c, November 8). [Letter to Deputy Director of Timber Supply on behalf of Officer-in-Charge Turner]. Dominions Office, Original Correspondence, 1943–46 (DO 35, Reel B 5021 /745, File N.264/52–213). The Rooms Provincial Archives, St. John's, NL, Canada.

Hart, A. E (1942a, February 26). [Letter to Geoffrey Shakespeare, Dominions Office]. Dominions Office, Original Correspondence, 1943–46 (DO 35, Reel B 5021 /745, File N.264/52–213). The Rooms Provincial Archives, St. John's, NL, Canada.

Hart, A. E (1942b, September 22). [Memorandum to General Carrington, Assistant Director, Home Grown Timber Production]. Newfoundland Forestry Unit in Scotland: welfare arrangements (AVIA 22 1352). The National Archives, London, UK.

Hart, A. E. (1942c, October 10). Newfoundland Forestry Unit: Use of official vehicles for recreational and church purposes [List of regulations]. Newfoundland Forestry Unit in Scotland: welfare arrangements (AVIA 22 1352). The National Archives, London, UK.

Hart, A. E. (1945, July 2). [Notes for file. Dominions Office]. Dominions Office, Original Correspondence, 1943–46 (DO 35, Reel B 5212 /1156, File N.718/1–15). The Rooms Provincial Archives, St. John's, NL, Canada.

Hart, P. J. (2010). The white feather campaign: A struggle with masculinity during World War I. *Inquiries Journal*, 2(2). http://www.inquiriesjournal.com/a?id=151

Hickey, T., & Puddester, J. (1938, July 25). Correspondence between unemployed and government. *The Evening Telegram*, 5. https://collections.mun.ca/digital/collection/telegram21/id/23769/rec/13

Higgins, J. (2007a). *Agriculture and the Commission of Government*. Heritage Newfoundland and Labrador. https://www.heritage.nf.ca/articles/politics/agriculture-commission-government.php

Higgins, J. (2007b). *Home defence*. Heritage Newfoundland and Labrador. https://www.heritage.nf.ca/articles/politics/home-defence-wwii.php

Higgins, J. (2007c). *Newfoundlanders and Labradorians in WW II*. Heritage Newfoundland and Labrador. https://www.heritage.nf.ca/articles/politics/newfoundlanders-labradorians-wwii.php

Higgins, J. (2015). *Newfoundland Ranger Force*. Heritage Newfoundland and Labrador. https://www.heritage.nf.ca/articles/politics/newfoundland-rangers.php

High, S. (2003, Spring). Working for Uncle Sam: The "comings" and "goings" of Newfoundland base construction labour, 1940–1945. *Acadiensis, 32*(2), 84–107. https://id.erudit.org/iderudit/acad32_2art04

Hill-billies come to Scotland. (1940, February 29). *The Evening Telegram*, 19. https://collections.mun.ca/digital/collection/telegram21/id/32268/rec/57

Hiller, J. K. (1997a). *British policy and Newfoundland, 1942–1946*. Heritage Newfoundland and Labrador. https://www.heritage.nf.ca/articles/politics/british-policy-newfoundland.php

Hiller, J. K. (1997b). *The Newfoundland National Convention, 1946–1948*. Heritage Newfoundland and Labrador. https://www.heritage.nf.ca/articles/politics/newfoundland-national-convention.php

Hiller, J. K. (2002). *The Newfoundland Royal Commission, 1933 (the Amulree Commission)*. Heritage Newfoundland and Labrador. https://www.heritage.nf.ca/articles/politics/amulree-commission.php

Hiller, J. K., & O'Brien, M. (2019). Newfoundland. In U. Daniel, P. Gatrell, O. Janz, H. Jones, J. Keene, A. Kramer, & B. Nasson (Eds.), *1914–1918-online: International encyclopedia of the First World War*. Freie Universität Berlin. https://doi.org/10.15463/ie1418.11346

Hodgsen, R. M. (1942, September 4). The boys from Newfoundland. *The Western Star*, 1. https://collections.mun.ca/digital/collection/westernstar/id/13462/rec/156

Home Bred. (1940, March 2). Work for both [Letter to the editor]. *Newcastle Journal*, 4.

Home Grown Timber Production Department. (1941, March 17). *Minutes of meeting at the Dominions Office with representatives of the Newfoundland Government on matters relating to the unit.* Dominions Office, Original Correspondence, 1943–46 (DO 35, Reel B 5021 /745, File N.264/52–213). The Rooms Provincial Archives, St. John's, NL, Canada.

Homicide charge deserted. (1940, September 18). *The Evening Telegraph* (Dundee), 3.

Hubbard, A. R. (1940, July 15). [Memo to Officer-in-Charge Turner re particulars of the death of Max Hawkins]. Dominions Office, Original Correspondence, 1943–46 (DO 35, Reel B 5021 /745, File N.264/52–213). The Rooms Provincial Archives, St. John's, NL, Canada.

Hyslop, B. (2001). *Journeys through time: Clarenville—hub of the east coast.* Town of Clarenville.

Increase in pay for foresters. (1943, February 12). *The Western Star*, 7. https://collections.mun.ca/digital/collection/westernstar/id/13779/rec/234

James, R. H. (1940, September 25). [Letter to Jack Turner re Death of Maxwell Hawkins (Case of Francis Lionel Harrington: Inverness High Court)]. Dominions Office, Original Correspondence, 1943–46 (DO 35, Reel B 5021 /745, File N.264/52–213). The Rooms Provincial Archives, St. John's, NL, Canada.

Janes, H. C. (1921, November 2). [Letter to Lieutenant M. C. Nugent]. The Great War Military Service Files Database. The Rooms Provincial Archives, St. John's, NL, Canada. https://therooms.ca/sites/default/files/rnr-database/import/nugent_michael_joseph_0-209_or_428.pdf

Jarratt, M. (2016). *Letters from Beauly: Pat Hennessy and the Canadian Forestry Corps in Scotland, 1940–1945.* Goose Lane.

Jeddore, J. N. (2001, May 21). *Letter from John Nick Jeddore to C. Pike.* Newfoundland Overseas Forestry Unit 1939–1945. http://www.nofu.ca/submitted_items/john_n_jeddore.html

Jeddore, J. N. (2015). *Moccasin tracks: A memoir of Mi'kmaw life in Newfoundland.* Memorial University Press.

Jenkinson, A. J. (1943, February16). [Letter to H. B. Acton, Home Grown Timber Production, Ministry of Supply]. Newfoundland Forestry Unit in Scotland: welfare arrangements (AVIA 22 1352). The National Archives, London, UK.

J. G. (1945, March). Letter to the editor. *Newfoundland Foresters' Bulletin, 5,* 2.

J. T. (1945, March). Letter to the editor. *Newfoundland Foresters' Bulletin, 5,* 1–2.

Kelly, U. A. (2014). *Mentioned in song: Song traditions of the loggers of Newfoundland and Labrador* [audio disc and booklet]. Research Centre for the Study of Music, Media and Place, Memorial University.

Kelly, U. A. (2018). The occupational songs, recitations, poems, and stories of the woods workers of Newfoundland and Labrador: An introduction. In U. A. Kelly & M. C. Forsyth (Eds.), *The music of our burnished axes: Songs and stories of the woods workers of Newfoundland and Labrador* (pp. 39–55). Memorial University Press.

Kelly, U. A. (2020). *The foresters' scribe: Remembering the Newfoundland Forestry Companies through the letters of Regimental Quartermaster Sergeant John A. Barrett.* Memorial University Press.

Kelly, U. A. (2021a). Linerboard mill woods operations. *Them Days, 45*(2), 44–48.

Kelly, U. A. (2021b) (Ed.). Songs and stories of the woods workers of Labrador [Themed issue]. *Them Days, 45*(2), 2–80.

Kelly, U. A. (2022). Revitalizing the woods-song tradition of Newfoundland and Labrador. *Doyle's Almanac of Newfoundland and Labrador 2023, 8,* 24–30.

Kelly, U. A., & Forsyth, M. C. (2018). *The music of our burnished axes: Songs and stories of the woods workers of Newfoundland and Labrador.* Memorial University Press.

Kelly, U. A., & Forsyth, M. C. (2021). *The music of our burnished axes: New recordings of woods songs and recitations from Newfoundland and Labrador* [audio disc and booklet]. Research Centre for the Study of Music, Media and Place, Memorial University.

Kieley, M. C. (1985). Memorates of Michael Collier [unpublished manuscript]. Folklore and Language Archives, 86–006, Memorial University, St. John's, NL, Canada.

King, D. (1999, September). *In the name of liberalism: Illiberal social policy in the USA and Britain*. Oxford University Press. https://doi.org/10.1093/0198296290.001.0001

Kitchen, J. (2005). *By the sweat of my brow*. Author.

La Fédération des communautés francophones et acadienne du Canada. (2009). *Francophone community profile of Newfoundland and Labrador*. https://www.francotnl.ca/FichiersUpload/Documents/20090910Profils_terre_neuve_et_labrador_en.pdf

Labrador's veterans. (2020). *Them Days, 44*(1), 1–192.

Land clearing expedited by brush rake. (1951, December). *Newfoundland Government Bulletin*, 7–8. https://collections.mun.ca/digital/collection/nlbulletin/id/1979/rec/1

Laracy, T. (1945, April). Letter to the editor. *Newfoundland Foresters' Bulletin*, 6, 1–2.

The last lumberjack. (2006, January 14). *The Herald*. https://tinyurl.com/2xm5ekpr

Left car in dark—accident follows. (1941, January 31). *The Hawick News*, 3.

Legion's provincial command passes resolutions, elects officers. (1950, October 24). *The Western Star*, 1, 5. https://collections.mun.ca/digital/collection/westernstar/id/23349/rec/236

Legion to back foresters demands for war pensions. (1958, November 15). *The Western Star*, 1. https://collections.mun.ca/digital/collection/westernstar/id/51336/rec/208

Lenanton, G. L. (1942a, March 10). [Cover letter to Turner reports September 1941 to January 1942]. Dominions Office, Original Correspondence, 1943–46 (DO 35, Reel B 2511 /1155, File N.664/4–65). The Rooms Provincial Archives, St. John's, NL, Canada.

Lenanton, G. L. (1942b, October 23). [Memorandum to C. J. Pyke, Raw Materials, Ministry of Supply]. Newfoundland Forestry Unit in Scotland: welfare arrangements (AVIA 22 1352). The National Archives, London, UK.

Letter from a local boy with the foresters. (1940, February 27). *The Daily News*, 3. https://collections.mun.ca/digital/collection/dailynews/id/119923/rec/2

Ley Smith, G. E. (1941, April 4). Lumberjacks wrangle way into services. *Aberdeen Press and Journal*, 4.

Lintott, H. J. B. (1942, November 11). [Letter to C. J. Pyke, Raw Materials, Ministry of Supply]. Newfoundland Forestry Unit in Scotland: welfare arrangements (AVIA 22 1352). The National Archives, London, UK.

Lloyd, A. H. (1944). The operations of military forestry units in France and Great Britain, 1939–1944. *Empire Forestry Journal*, *23*(1), 10–19. https://www.jstor.org/stable/42594426

Lloyd, A. H. (1945). Timber supplies for the Allied Expeditionary Force, 1944–45. *Empire Forestry Journal*, *24*(2), 141–149. http://www.jstor.com/stable/42594376

Lloyd George, G. (1942, September 8). [Letter to Clement Attlee, Dominions Office]. Newfoundland Forestry Unit in Scotland: welfare arrangements (AVIA 22 1352). The National Archives, London, UK.

Local foresters organize, seek help from federal authorities. (1960, April 4). *The Evening Telegram*, 3. https://collections.mun.ca/digital/collection/telegram21/id/154837/rec/1

Local made stoves for forestry battalion. (1940, January 12). *The Evening Telegram*, 4. https://collections.mun.ca/digital/collection/telegram21/id/29759/rec/61

Loggers invest savings. (1943, March 26). *The Daily News*, 3. https://collections.mun.ca/digital/collection/dailynews/id/107435/rec/233

The loggers of the North. (1940, November 2). *The Evening Chronicle* (Newcastle), 6.

Loggers to work here. (1939, December 19). *The Northern Whig*, 6.

Loggers urged to stay on job. (1940, April 24). *Press and Journal* (Aberdeen), 6.

Loggers wanted at Grand Falls and Corner Brook. (1942, June 27). *The Daily News*, 3. https://collections.mun.ca/digital/collection/dailynews/id/112236/rec/3

London letter: Wartime thinking. (1944, February 14). *Birmingham Gazette*, 2.

Long, G. (1999). *Suspended state: Newfoundland before Canada*. Breakwater Books.

Long, G. M. (1952). *Australia in the war of 1939–1945: To Benghazi* (Vol. 1). Australian War Memorial.

Low forestry production: Comptroller's comment. (1941, April 24). *Press and Journal* (Aberdeen), 4.

Lumberjack shot on road. (1940, July 16). *The Courier and Advertiser* (Dundee), 3

Lumberjacks. (1940, February 21). *The Hawick Express*, 4.

Lumberjacks at play. (1942, September 7). *Evening Express*, 5.

Lumberjacks fined. (1940, October 8). *The Berwickshire News*, 3.

Lumberjacks in Scotland. (1940, March 21). *Weekly News* (Runcorn), 6.

Lumberjacks: The Newfoundlanders in Scotland. (1941, April 4). *The Scotsman*, 7.

Lumberjack's wife. (1941, December 28). A wife's appeal [Letter to the editor]. *The Sunday Post*, 5.

Lumberjills. (1942, July 3). *The Western Star*, 2. https://collections.mun.ca/digital/collection/westernstar/id/13366/rec/2

Lumberjills beat lumberjacks. (1942, July 4). *The Kingston Times*, 4.

Lumberjills for Newfoundland? (1942, June 19). *The Western Star*, 2. https://collections.mun.ca/digital/collection/westernstar/id/13340/rec/4

Lumberman fined £10. (1940, April 17). *Dumfries and Galloway Standard and Advertiser*, 8.

Lumbermen in court. (1941, September 11). *Kirriemuir Free Press and Angus Advertiser*, 1.

Lumbermen like the Scots, but not our climate. (1941, April). *Daily Record and Mail*, 6

Lumbermen love our jam, hate our damp cold. (1939, December 23). *News Chronicle*, 8.

MacDonald, G. (1946, April). Message from Governor of Newfoundland. *Newfoundland Foresters' Bulletin*, *18*, 1. http://nofu.ca/submitted_items/skenway_bulletin_p1.html

Machtig, E. (1942, July 15). [Letter to Governor Walwyn with Enclosure: Newfoundland Forestry Unit]. Governor's Office – Local and

Miscellaneous Correspondence 1942 (GN 1/3/A, Box 242, File 2). The Rooms Provincial Archives, St. John's, NL, Canada.

MacKenzie, D. (2010). The Indian Act and the Aboriginal Peoples of Newfoundland at the time of Confederation. *Newfoundland [and Labrador] Studies*, 25(2), 161–181. https://www.erudit.org/en/journals/nflds/2010-v25-n2-nflds25_2/nflds25_2art01/

The man in a man's job: Ogden's St. Bruno. (1940, July 22). *Leicester Evening Mail*, 7.

March, A. H. (1964, August 4). [Letter to Premier Smallwood]. J. R. Smallwood Papers (Collection 075, Box 191, File 3.10.032). Archives and Special Collections, Memorial University of Newfoundland, St. John's, Canada.

Marshall, F. W. (1951, September 11). Major F. W. Marshall addresses Canadian Legion conference. *The Daily News*, 4. https://collections.mun.ca/digital/collection/dailynews/id/69788/rec/77

Martin, W. B. W. (1994). *Random Islanders On Guard*. Creative Publishing.

Mather, A. S. (1978, July). Patterns of afforestation in Britain since 1945. *Geography*, 63(3), 157–166. https://www.jstor.org/stable/40568942

Matthews, F. G. (1961a, April 15). The Legion corner. *The Western Star*, 6, 16. https://collections.mun.ca/digital/collection/westernstar/id/62240/rec/2

Matthews, F. G. (1961b, December 2). The Legion corner. *The Western Star*, 5. https://collections.mun.ca/digital/collection/westernstar/id/64285/rec/33

Matthews, F. G. (1962, December 8). The Legion corner. *The Western Star*, 5. https://collections.mun.ca/digital/collection/westernstar/id/69203/rec/26

Matthews, F. G. (1963, February 16). The Legion corner. *The Western Star*, 5, 8. https://collections.mun.ca/digital/collection/westernstar/id/70016/rec/1

Matthews, F. G. (1965a, April 24). The Legion corner. *The Western Star*, 12. https://collections.mun.ca/digital/collection/westernstar/id/80489/rec/1

Matthews, F. G. (1965b, November 20). The Legion corner, *The Western Star*, 7, 16. https://collections.mun.ca/digital/collection/westernstar/id/84468/rec/2

Matthews, F. G. (1966, June 11). The Legion corner. *The Western Star*, 16. https://collections.mun.ca/digital/collection/westernstar/id/85880/rec/77

Maxton, J. (1944, February 11). "Dominion services." UK Parliament. House of Commons. *Debates Hansard*, Vol. 396(2021). https://api.parliament.uk/historic-hansard/commons/1944/feb/11/dominion-services

McCann, P. (1987, October). The educational policy of the Commission of Government. *Newfoundland and Labrador Studies*, 3(2), 201–215. https://journals.lib.unb.ca/index.php/NFLDS/article/view/591

McGrath, J. (1962, February 14). "Civilian War Pensions." Canada. Parliament. House of Commons. *Debates* 1. 24th Parliament, 5th session. Canadian Parliamentary Historical Resources: https://n2t.net/ark:/69429/m0m03xs5ns0j

McGrath, J., Furlong, J., Hackett, W., Dicks, W., Rogers, R., Purcell, S., Evans, T., Dillon, M., Fitzgerald, M., Connolly, J., & Foley, D. (1932, February 8). Consider new dole scale unreasonable. *The Daily News*, 4. https://collections.mun.ca/digital/collection/dailynews/id/147012/rec/8

McIvor, A. (2018). Rebuilding 'real men': Work and working-class male civilian bodies in wartime. In L. Robb & J. Pattinson (Eds.), *Men, masculinities and male culture in the Second World War* (pp. 121–144). Palgrave Macmillan.

McKelvey, P. (2001). New Zealand foresters at war. *NZ Journal of Forestry*, 45(4), 34–39. https://nzif.org.nz/nzif-journal/publications/article/21889

McKenzie-Smith, G. R. (2018). *The unit guide: The Australian army 1939–1945*. Vol. 3. Big Sky Publishing.

McNeil, P. (1993). 'Put your best face forward': The impact of the Second World War on British dress. *Journal of Design History*, 6(4), 283–299. https://www.jstor.org/stable/1316093

Men proud to serve Britain need aid. (1940, May 5). *The People*, 7.

Men wanted [Public notice]. (1941, July 25). *The Daily News*, 1. https://collections.mun.ca/digital/collection/dailynews/id/116614/rec/189

Men wanted for the Newfoundland Overseas Forestry Unit. (1941, May 7). *The Western Star*, 8. https://collections.mun.ca/digital/collection/westernstar/id/12756/rec/23

Mercer, I. (1950a, April 29). [Letter to Premier Smallwood]. J. R. Smallwood Papers, Collection 075, Box 144, File 3.03.012). Archives and Special Collections, Memorial University of Newfoundland, St. John's, Canada.

Mercer, I. (1950b, July 3). [Letter to Premier Smallwood]. J. R. Smallwood Papers, Collection 075, Box 191, File 3.10.031). Archives and Special Collections, Memorial University of Newfoundland, St. John's, Canada.

Mishaud, E. (1998, May 26). Norman Gabriel: A living legend. *The Georgian* (Stephenville, NL), 3.

Modsell, H. M. (1939, December 6). Memo. In Relieving officers' notice [Editorial]. *The Evening Telegram*, 4. https://collections.mun.ca/digital/collection/telegram21/id/28293/rec/10

More lumberjacks for Britain. (1940, January 19). *The Evening Telegraph* (Dundee), 3.

Morris, E. P. (1917a, March 14). The Prime Minister writes from London. *The Daily News*, 4. https://collections.mun.ca/digital/collection/dailynews/id/233425/rec/3

Morris, E. P. (1917b, July 2). First anniversary of the battle of Beaumont-Hamel [Public address]. *The Evening Advocate*, 6. https://collections.mun.ca/digital/collection/eveadvocate/id/437/rec/5

Mr. Attlee with lumbermen. (1942, September 7). *Press and Journal* (Aberdeen), 3.

Munro, J. (2011a). A good place to start a woods business (part one). *Them Days*, *35*(1), 50–63.

Munro, J. (2011b). A good place to start a woods business (part two). *Them Days*, *35*(2), 46–62.

Neary, P. (1969). Democracy in Newfoundland: A comment. *Journal of Canadian Studies*, *4*(1), 37–45. https://www.muse.jhu.edu/article/675967

Neary, P. (1973). *The political economy of Newfoundland, 1929–1972*. Copp Clark.

Neary, P. (1983). Newfoundland's union with Canada, 1949: Conspiracy or choice? *Acadiensis*,*12*(2), 110–119. https://journals.lib.unb.ca/index.php/Acadiensis/article/view/11591

Neary, P. (1984). Clement Attlee's visit to Newfoundland, September 1942. *Acadiensis*, *13*(2), 101–109. https://journals.lib.unb.ca/index.php/Acadiensis/article/view/12228

Neary, P. (1985). The Bradley report on logging operations in Newfoundland, 1934: A suppressed document. *Labour/Le Travail*, *16*, 193–232. https://www.lltjournal.ca/index.php/llt/article/view/2477

Neary, P. (1988). *Newfoundland in the North Atlantic world 1929 to 1949*. McGill-Queen's University Press.

Neary, P. (1993a). Ebb and flow: Citizenship in Newfoundland 1929–1949. In W. Kaplan (Ed.), *Belonging: The meaning and future of Canadian citizenship* (pp. 79–103). McGill-Queen's University Press.

Neary, P. (1993b). *How Newfoundland veterans became Canadian veterans: A study in bureaucracy and benefit*. (Western University Political Economy Research Group, Papers in Political Economy, No. 30). https://uwo.scholaris.ca/server/api/core/bitstreams/99861f7a-295b-4da8-aa5c-857e8efe4850/content

Neary, P. (2011a). The Commission of Government on reconstruction, December 1936. *Newfoundland and Labrador Studies, 26*(2), 241–288. https://journals.lib.unb.ca/index.php/NFLDS/article/view/21035/24265

Neary, P. (2011b). *On to civvy street: Canada's rehabilitation program for veterans of the Second World War*. McGill-Queens University Press.

Neary, P. (2012). P. A. Clutterbuck on Morley Richards and the record of the Commission of Government, 1939. *Newfoundland and Labrador Studies, 27*(1), 80–98. https://journals.lib.unb.ca/index.php/NFLDS/article/view/21047

Neary, P. (2017). High Commissioner J. J. S. Garner on Joey Smallwood versus John Diefenbaker, 1959. *Newfoundland and Labrador Studies, 32*(1). https://journals.lib.unb.ca/index.php/NFLDS/article/view/26103

New defence corps recruitment. (1942, May 16). *Press and Journal* (Aberdeen), 3.

Newfoundland Commission of Enquiry on the Logging Industry. (1960). *Report*. https://collections.mun.ca/digital/collection/cns2/id/163204/rec/9

Newfoundland Commission of Government (NCG). (1945, April 7). [Telegram to Dominions Office]. Dominions Office, Original Correspondence, 1943–46 (DO 35, Reel B 5212 /1156, File N.718/1–15). The Rooms Provincial Archives, St. John's, NL, Canada.

Newfoundlanders at university. (1946, March 5). *Press and Journal* (Aberdeen), 3.

Newfoundlanders become honorary members of the Regimental Association of the Queen's Own Cameron Highlanders. (1944, December 1).

Newfoundland Foresters' Bulletin, 2, 3. Photographs and Cuttings Related to the Newfoundland Overseas Forestry Unit Collected by Tom Hall (PD 19). Aberdeen City and Aberdeenshire Archives, Aberdeen, Scotland.

Newfoundlanders return home. (1940, August 26). *The Evening Telegram*, 6. https://collections.mun.ca/digital/collection/telegram21/id/30772/rec/1

Newfoundlanders win esteem abroad. (1943, May 22). *The Western Star*, 14. https://collections.mun.ca/digital/collection/westernstar/id/14000/rec/15

Newfoundland loggers. (1940, May 11). *Sussex and Surrey Courier*, 6.

Newfoundland loggers' badge. (1940, June 29). *The Journal of Commerce and Shipping Telegraph* (Liverpool), 4.

Newfoundland loggers clearing Scottish forests. (1941, March 1). *The Weekly Scotsman*, 1.

Newfoundland loggers helping to beat submarine menace. (1941, May 3). *The Newfoundland Weekly*, 1(22), 4, 8. https://collections.mun.ca/digital/collection/nfldweekly/id/2199/rec/249

Newfoundland lumberjacks in Britain. (1939, December 20). *The Nottingham Evening Post*, 5.

Newfoundland Statistics Agency (1988). *Historical statistics of Newfoundland and Labrador.* https://www.stats.gov.nl.ca/publications/Historical/PDF/Historical_Statistics_of_Newfoundland_and_Labrador_V2_N5_1988.pdf

News about people: Bruce Fraser. (1947, February 7). *The Western Star*, 3. https://collections.mun.ca/digital/collection/westernstar/id/17568/rec/21

News budget from the 59th Heavy Regiment (1944, April 11). *The Daily News*, 9. https://collections.mun.ca/digital/collection/dailynews/id/103841/rec/2

Nfld. foresters in the Highlands. (1917, October 20). *The Evening Telegram*, 4. https://collections.mun.ca/digital/collection/telegram19/id/12299/rec/2

Nfld. Forestry Unit [Editorial]. (1939, November 20). *The Evening Telegram*, 6. https://collections.mun.ca/digital/collection/telegram21/id/27984/rec/7

Nfld. Forestry Unit attends study courses. (1944, June 17). *The Daily News*, 4. https://collections.mun.ca/digital/collection/dailynews/id/104973/rec/68

Nfld. Forestry Unit at work in Argyllshire. (1940, November 30). *The Daily News*, 3. https://collections.mun.ca/digital/collection/dailynews/id/122664/rec/239

Nicholson, G. W. L. (1969). *More fighting Newfoundlanders: A history of Newfoundland's fighting forces in the Second World War*. Government of Newfoundland and Labrador.

Noel, S. J. R. (1971). *Politics in Newfoundland*. University of Toronto Press.

NOFA (Newfoundland Overseas Foresters' Association). (1945). *Timber!!* Newfoundland Overseas Foresters' Association. https://collections.mun.ca/digital/collection/cns2/id/41706/rec/26

NOFA (Newfoundland Overseas Foresters' Association). (1949). *Brief submitted on behalf of the Newfoundland Overseas Forestry Unit*. https://collections.mun.ca/digital/collection/cns_tools/id/143901/rec/7

NOFU Superintendents. (1942, May 26). [Untitled document]. Dominions Office, Original Correspondence, 1943–46 (DO 35, Reel B 2511 /1155, File N.664/4–65). The Rooms Provincial Archives, St. John's. NL, Canada.

No Oil in Our Lamps and No Butter on Our Bread. (1934, August 17). Complaints of lumberman [Letter to the editor]. *The Evening Telegram*, 6. https://collections.mun.ca/digital/collection/telegram21/id/6060/rec/2

Notes on executive meetings. (1944, November 1). *Newfoundland Foresters' Bulletin*, *1*, 3. Photographs and Cuttings Related to the Newfoundland Overseas Forestry Unit Collected by Tom Hall (PD 19). Aberdeen City and Aberdeenshire Archives, Aberdeen, Scotland.

Not heard of bigamy. (1942, December 23). *The Northern Daily Mail* (Hartlepool), 4.

Notice. (1945, November 23). *The Western Star*, 4. https://collections.mun.ca/digital/collection/westernstar/id/16175/rec/5

NRC (Newfoundland Royal Commission) report. (1933). His Majesty's Stationery Office. https://collections.mun.ca/digital/collection/cns/id/19703/rec/26

Obituaries: Mrs. B. L. Bode. (1997, March 12). *Post and Times* (Leek), 34.

O'Brien, T. (Producer). (2005) *Newfoundland foresters' contribution to WW II* [Video]. CBC (St. John's). https://www.cbc.ca/news/canada/

newfoundland-labrador/newfoundland-foresters-contribution-to-ww-ii-1.2667654

Obstacles to local manufactures [Editorial]. (1945, September 11). *The Evening Telegram*, 6. https://collections.mun.ca/digital/collection/telegram21/id/51875/rec/1

Old world can help the new. (1944, June 6). *Press and Journal* (Aberdeen), 4.

Oosthoek, J. (2013). *Conquering the Highlands: A history of the afforestation of the Scottish uplands*. ANU E Press.

Opera on the Avalon. (2024, March 30). *1949* [Video]. YouTube. https://www.youtube.com/watch?v=MMBFiimf7vk

Operations. (n.d.). [Unpublished manuscript]. Photographs and Cuttings Related to the Newfoundland Overseas Forestry Unit Collected by Tom Hall (PD 19 1). Aberdeen City and Aberdeenshire Archives, Aberdeen, Scotland.

Ormiston Kelly, M. (2002, September). *Newfoundland Overseas Foresters Association annual meeting and reunion keynote address*. Newfoundland Overseas Forestry Unit 1939–1945. http://www.nofu.ca/submitted_items/mulliadh_kelly.html

Orr, J. R. (1942, January 29). [Letter to Dorothy Thompson, Empire Societies War Hospitality Committee]. Dominions Office, Original Correspondence, 1943–46 (DO 35, Reel B 5021 /745, File N.264/52–213). The Rooms Provincial Archives, St. John's, NL, Canada.

Over the teacups: Lumberjacks. (1940, June 15). *The Galway Observer*, 4.

Overton, J. (2001). Poverty, dependence and self-reliance: Politics, Newfoundland history and the Amulree report. In G. Fizzard (Ed.), *Amulree's legacy: Truth, lies and consequences symposium, March 2000* (pp. 47–68.). Newfoundland Historical Society. https://collections.mun.ca/digital/collection/cns2/id/21752/rec/4

Oxford University Press (n.d.). Forester. In *Oxford English Dictionary*. Retrieved February 22, 2024, from https:/www.oed.com

Oxford University Press (n.d.). Logger. In *Oxford English Dictionary*. Retrieved February 22, 2024, from https:/www.oed.com

Oxford University Press (n.d.). Lumberjack. In *Oxford English Dictionary*. Retrieved February 22, 2024, from https:/www.oed.com

Paid for doing no work. (1941, April 24). *Birmingham Daily Gazette*, 4.

Paine, R. (1986, Fall). The mastery of Smallwood and interlocutory rhetoric. *Newfoundland [and Labrador] Studies*, 2(2), 191–211. https://www.erudit.org/en/journals/nflds/1986-v2-n2-nflds_2_2/nflds2_2art02/

Passingham, W. J. (1941, February 8). We visit a Scottish lumber camp. *Illustrated*, 1–6. Reprinted in *The Daily News*, December 31, 1942, 59. https://collections.mun.ca/digital/collection/dailynews/id/114512/rec/14

Patriotic Association of the Women of Newfoundland (WPA). (1943). *Annual report* 1942/43. https://collections.mun.ca/digital/collection/cns2/id/29588/rec/1

Paul, I. (2000). Ignatius Paul. In C. Coish (Ed.), *Stories of the Mi'kmaq* (pp. 39–44). St. John's: College of the North Atlantic. https://web.archive.org/web/20150324192147/http://en.copian.ca/library/learning/mikmaq/mikmaq.pdf

Peat, C. U. (1942a, March 21). [Letter to Paul Emrys-Evans]. Dominions Office, Original Correspondence, 1943–46 (DO 35, Reel B 5021 /745, File N.264/52–213). The Rooms Provincial Archives, St. John's, NL, Canada.

Peat, C. U. (1942b, June 8). [Letter to Clement Attlee with enclosure: Peat report on NOFU camps]. Newfoundland Forestry Unit in Scotland: welfare arrangements (AVIA 22 1352). The National Archives, London, UK.

Peat, C. U. (1942c, June 30). [Letter to Paul Emrys-Evans with enclosure: Note of agreement between Ministry of Supply and Dominions Office]. Newfoundland Forestry Unit in Scotland: welfare arrangements (AVIA 22 1352). The National Archives, London, UK.

Peniston-Bird, C. (2018). Commemorating invisible men: Reserved occupations in bronze and stone. In L. Robb & J. Pattinson (Eds.), *Men, masculinities and male culture in the Second World War* (pp. 189–214). Palgrave Macmillan.

Penson, J. H. (1941a, March 7). *Visit by Mr. Penson to Newfoundland Forestry Unit* [Report]. Dominions Office, Original Correspondence, 1943–46 (DO 35, Reel B 5021 /745, File N.264/52–213). The Rooms Provincial Archives, St. John's, NL, Canada.

Penson, J. H. (1941b, May 1). Nfld. loggers play big role in war effort. *The Daily News*, 3, 8. https://collections.mun.ca/digital/collection/dailynews/id/115769/rec/2

Pike, S. (n.d.). *Excerpts from a forester's autograph book*. Newfoundland Overseas Forestry Unit 1939–1945. http://nofu.ca/autograph_book_main_page.html

Pitwood. (1940, February 15). Timber felling in Scotland [Letter to the editor]. *The Scotsman*, 9.

Placentia branch foresters unit sends letter to prime minister. (1962, October 17). *The Daily News*, 5. https://collections.mun.ca/digital/collection/dailynews/id/40890/rec/1

Placentia forms forestry unity. (1960, March 3). *The Evening Telegram*, 5. https://collections.mun.ca/digital/collection/telegram21/id/153596/rec/2

Pollard, G. (1940, January 10). Newfoundlanders in England. *The Western Star*, 1, 6. https://collections.mun.ca/digital/collection/westernstar/id/12093/rec/1

Porter, B. (1994). Other things to do: An interview with Gregory Power. *TickleAce*, *28*, 14–30.

Powell, J. A. (1947, September 30). [Letter to Secretary of Finance]. Department of Finance Correspondence: Newfoundland Forestry Unit (GN 158.176). The Rooms Provincial Archives, St. John's, NL, Canada.

Powers, B. (2000). The legacy of Captain Jack Turner. *The Forestry Chronicle*, *76*(6), 828–829. https://doi.org/10.5558/tfc76820-6

Provincial Archives of NL and Memorial University Television (Producers). (1986). *World War I and II films: Newfoundlanders at war* [Video]. Media and Data Centre, Memorial University of Newfoundland. https://collections.mun.ca/digital/collection/extension/id/900/rec/1

Provision for Forestry Unit case accident. (1942, March 19). *The Daily News*, 3. https://collections.mun.ca/digital/collection/dailynews/id/111456/rec/450

Public notice. (1939, November 29). *The Newfoundland Gazette*, *15*(48), 1. https://collections.mun.ca/digital/collection/nl_gazette/id/3564/rec/1

Public notice. (1942, October 20). *The Newfoundland Gazette*, *18*(42), 1. https://collections.mun.ca/digital/collection/nl_gazette/id/4700/rec/5

Public notice: Defence medal. (1946, September 27). *The Western Star*, 15. https://collections.mun.ca/digital/collection/westernstar/id/17118/rec/1

Published by authority. (1943, March 23). *The Newfoundland Gazette*, 19(12), 1. https://collections.mun.ca/digital/collection/nl_gazette/id/4799/rec/6

Puddester, J. C. (1939, November 22). [Telegram to Secretary of State for Dominion Affairs]. Governor's Office – Local and Miscellaneous Correspondence 1936—Forestry Recruitment (GN 1/3/A, Box 187, File 118). The Rooms Provincial Archives, St. John's, NL, Canada.

Quigley, E. (2019, February 1). *Scotland's debt to forgotten Belize lumberjacks*. BBC. https://www.bbc.com/news/uk-scotland-47049573

Record of the year at home in review. (1944, January 4). *Observers' Weekly*, 9–12. https://collections.mun.ca/digital/collection/observersweek/id/10802/rec/3

Reddigan, J. (2009, May 16). My father, Newfoundland lumberjack. *The Telegram*, 8.

Relieving officers' notice [Editorial]. (1939, December 6). *The Evening Telegram*, 4. https://collections.mun.ca/digital/collection/telegram21/id/28293/rec/10

Reubens, B. G. (1945). Unemployment in war-time Britain. *The Quarterly Journal of Economics*, 59(2), 206–236. https://doi.org/10.2307/1884826

Review and prospect. (1939, December 30). *The Bay Roberts Guardian*, 4. https://collections.mun.ca/digital/collection/bayroberts/id/5620/rec/210

Reyburn, W. (1942, August 28). Scotland suits them. *The Western Star*, 1. https://collections.mun.ca/digital/collection/westernstar/id/13449/rec/10

Richards, M. (1941, March 21). First colony goes to war. *The Scottish Daily Express*, 1–2.

Robb, L., & Pattinson, J. (2018). Becoming visible: Gendering the study of men at war. In L. Robb & J. Pattinson (Eds.), *Men, masculinities and male culture in the Second World War* (pp. 1–24). Palgrave Macmillan.

Roberts, C. (1998). Sally's Cove. In *Looking Back* (pp.66–67). Rocky Harbour NL Pioneers Club.

Roberts, E. (Edward). (2014). *A Blue Puttee at war: The memoir of Captain Sydney Frost, MC*. Flanker Press.

Robertson, A. (n.d.). *The Newfoundland connection*. https://collections.mun.ca/digital/collection/cns/id/146182/rec/61

Robinson, R. C. (1940a, March 26). [Letter to R. Ewbank, Dominions Office]. Dominions Office, Original Correspondence, 1943–46 (DO 35, Reel B 5020 /744, File N.264/30–50). The Rooms Provincial Archives, St. John's, NL, Canada.

Robinson, R. C. (1940b, March 29). [Letter to R. Ewbank, Dominions Office]. Dominions Office, Original Correspondence, 1943–46 (DO 35, Reel B 5020 /744, File N.264/30–50). The Rooms Provincial Archives, St. John's, NL, Canada.

Robinson, R. C. (1940c, June 21). [Covering note to Officer-in-Charge J. Turner's May 1940 report]. Commission of Government, Natural Resources—General Administration 1940 (GN 38.2, S2–1-17). The Rooms Provincial Archives, St. John's, NL, Canada.

Robinson, R. C. (1940d, July 11). [Covering note to Officer-in-Charge J. Turner's June 1940 report]. Commission of Government, Natural Resources—General Administration 1940 (GN 38.2, S2–1-17). The Rooms Provincial Archives, St. John's, NL, Canada.

Robinson, R. C. (1940e, October 15). [Note on the Newfoundland Forestry Unit from the Chair of the Forestry Commission to the Commission of Government of Newfoundland.] Dominions Office, Original Correspondence, 1943–46 (DO 35, Reel B 5021 /745, File N.264/52–213). The Rooms Provincial Archives, St. John's, NL, Canada.

Robinson, R. C. (1940f, November 28). [Covering note to Officer-in-Charge Turner's October 1940 report]. Dominions Office, Original Correspondence, 1943–46 (DO 35, Reel B 5021 /745, File N.264/52–213). The Rooms Provincial Archives, St. John's, NL, Canada.

Rose, S. O. (2003). *Which people's war? National identity and citizenship in Britain 1939–1945*. Oxford University Press.

Rouse, A. F. (1945, March 21). [Letter to St. John Chadwick, Dominions Office]. Dominions Office, Original Correspondence, 1943–46 (DO 35, Reel B 5212 /1156, File N.718/1–15). The Rooms Provincial Archives, St. John's, NL, Canada.

Royal Canadian Legion (2016). *Membership manual.* https://portal.legion.ca/docs/default-source/branch-and-command-resources/membership/membershipmanual_2016_e.pdf?sfvrsn=f5bb7a58_2

Ryan, A. (1949, July 25). [Letter to Premier Smallwood]. J. R. Smallwood Papers (Collection 075, Box 191, File 3.10.031). Archives and Special Collections, Memorial University of Newfoundland, St. John's, NL, Canada.

Sale of Labrador or confederation? (1932, August 17). *The Western Star,* 2. https://collections.mun.ca/digital/collection/westernstar/id/9210/rec/2

Saunders, G. (1960a, September 28). Today's foresters are highly trained college graduates. *The Evening Telegram,* 28. https://collections.mun.ca/digital/collection/telegram21/id/147884/rec/69

Saunders, G. (1960b, October 6). Col. Jack Turner built the foundation for today's work. *The Evening Telegram,* 21. https://collections.mun.ca/digital/collection/telegram21/id/148240/rec/4

Schoenfeld, M. H., & Whitney, A. L. (1942, Summer). Wartime methods of dealing with labor in Great Britain and the Dominions. *Law and Contemporary Problems,* 9(3), 522–543. https://scholarship.law.duke.edu/lcp/vol9/iss3/11

Scott, C. (2017). *Holding the home front: The Women's Land Army in the First World War.* Pen and Sword.

Scott, J. (1995). Oration honouring Dr. Thomas Vincent Curran. *The Gazette,* 27(19), 18. https://collections.mun.ca/digital/collection/mun_gazette/id/8788/rec/8

Scottish Half-Hour. (1944, May 17). *The Falkirk Herald,* 5.

Sears, J. L. (1942, July 15). [Letter to T. Fitzgerald, Home Grown Timber Production, Ministry of Supply]. Newfoundland Forestry Unit in Scotland: welfare arrangements (AVIA 22 1352). The National Archives, London, UK.

Seymour, E. E. (1976). *An illustrated history of Canadian labour 1800–1974.* Canadian Labour Congress.

Shakespeare, G. H. (1942, February 27). *Report on visit to Newfoundland Forestry camps, February 1942.* Newfoundland Forestry Unit in Scotland: welfare arrangements (AVIA 22 1352). The National Archives, London, UK.

Sheffield, R. S. (2018). Exploring the meaning of Indigenous military service during the Second World War in Australia, Canada, New Zealand and the United States. *Journal of Military and Strategic Studies, 19*(2), 71–89. https://jmss.org/article/view/62815/46847

Sheppard Alcock, J. (2015). Newfoundland and Labrador war brides. http://www.nlwarbrides.ca/

Shipping Advisory Committee. (1939, November 21). [Minutes of first meeting]. Commission of Government, Natural Resources, General Administration 1939–1942 (GN 38.2, S2–1-14A, File 6). The Rooms Provincial Archives, St. John's, NL, Canada.

Sider, G. (2003). *Between history and tomorrow: Making and breaking everyday life in rural Newfoundland*. Broadview Press.

Simmonds, L. H. (1917, April 13). Forestry battalion [Letter to the editor]. *The St. John's Daily Star*, 8. https://collections.mun.ca/digital/collection/daily_star/id/12577/rec/1

Six towns incorporated during 1945. (1946, January). *Newfoundland Government*, 17–20. https://collections.mun.ca/digital/collection/nlbulletin/id/1182/rec/2

Smallwood, J. R. (1950, July 3). [Letter to Minister Milton Gregg, Department of Veterans Affairs]. J. R. Smallwood Papers (Collection 075, Box 191, File 3.10.031). Archives and Special Collections, Memorial University of Newfoundland, St. John's, NL, Canada.

Sneddon, D. (2007). Newfoundlanders in a Highland forest during WW II. In T. Pollard & I. Banks (Eds.), *Scorched earth: Studies in the archaeology of conflict* (pp. 233–266). Brill. https://doi.org/10.1163/ej.9789004164482.i-330.95

Squibb, M. (2018, August 9). Scottish family meets with Newfoundland side of family for first time. *The Telegram*. https://www.saltwire.com/newfoundland-labrador/lifestyles/scottish-family-meets-with-newfoundland-side-of-family-for-first-time-232686/

SS Lady Rodney due to dock 7 p.m. Friday. (1946, July 25). *The Evening Telegram*, 3. https://collections.mun.ca/digital/collection/telegram21/id/54941/rec/5

Stapleton, G. A. (1981). *The function of personal experience narratives in the oral tradition of a war-bride* [unpublished manuscript]. Folklore and Language Archives, 81–497, Memorial University, St. John's, NL, Canada.

Stewart, A. (2010). The "bloody post office": The life and times of the Dominions Office. *Contemporary British History, 24*(1), 43–66. https://doi.org/10.1080/13619460903565390

Stewart, A. L. (1944). *Report of Mr. A. Lindsay Stewart on his lectures to members of the Newfoundland Forestry Unit in the Carr Bridge area.* Governor's Office – Local and Miscellaneous Correspondence 1944 (GN 1/3/A, Box 256, File 1 to 2). The Rooms Provincial Archives, St. John's, NL, Canada.

St. John Chadwick, G. W. (1942a, February 13). *Newfoundland Forestry Unit* [Report] Dominions Office, Original Correspondence, 1943–46 (DO 35, Reel B 5021 /745, File N.264/52–213). The Rooms Provincial Archives, St. John's, NL, Canada.

St. John Chadwick, G. W. (1942b, February 16). [Meeting note]. Dominions Office, Original Correspondence, 1943–46 (DO 35, Reel B 5021 /745, File N.264/52–213). The Rooms Provincial Archives, St. John's, NL, Canada.

St. John Chadwick, G. W. (1942c, May 26). [Report on NOFU]. Dominions Office, Original Correspondence, 1943–46 (DO 35, Reel B 5021 /745, File N.264/52–213). The Rooms Provincial Archives, St. John's, NL, Canada.

St. John Chadwick, G. W. (1942d, September 19). [Letter to A. E. Hart, NOFU]. Dominions Office, Original Correspondence, 1943–46 (DO 35, Reel B 5021 /745, File N.264/52–213). The Rooms Provincial Archives, St. John's, NL, Canada.

St. John Chadwick, G. W. (1942e, September 25). [Minutes]. Dominions Office, Original Correspondence, 1943–46 (DO 35, Reel B 5021 /745, File N.264/52–213). The Rooms Provincial Archives, St. John's, NL, Canada.

St. John Chadwick, G. W. (1942f, September 28). [Letter to Hugh N. Tait]. Dominions Office, Original Correspondence, 1943–46 (DO 35, Reel B 5021 /745, File N.264/52–213). The Rooms Provincial Archives, St. John's, NL, Canada.

St. John Chadwick, G. W. (1944, June 16). The Newfoundland Forestry Unit. *Newfoundland News*, 5–6.

Story, G. M., Kirwin, W. J., & Widdowson, J. D. A. (Eds.). (2006a). Dole. In *Dictionary of Newfoundland English* (2nd ed. with supplement, p. 679). University of Toronto Press.

Story, G. M., Kirwin, W. J., & Widdowson, J. D. A. (Eds.). (2006b). Newfie. In *Dictionary of Newfoundland English* (2nd ed. with supplement, p. 343). University of Toronto Press.

Strang, S. (2023, December 4). *Constable, Police War Reserve, Frederick Herbert Clarke.* For Evermore—Stories of the Fallen. https://www.cwgc.org/stories/stories/constable-police-war-reserve-frederick-herbert-clarke/

Stringer, E. (2011). *Hodge's Cove.* Eric Stringer.

Sub-committees. (1945, January 1). *Newfoundland Foresters' Bulletin, 3,* 2. Photographs and Cuttings Related to the Newfoundland Overseas Forestry Unit Collected by Tom Hall (PD 19). Aberdeen City and Aberdeenshire Archives, Aberdeen, Scotland.

Summerfield, P., & Peniston-Bird, C. (2000). Women in the firing line: The Home Guard and the defence of gender boundaries in Britain in the Second World War. *Women's History Review, 9*(2), 231–255. https://doi.org/10.1080/09612020000200250

Summerfield, P., & Peniston-Bird, C. (2007). *Contesting home defence: Men, women and the* Home Guard in the Second World War. Manchester University Press.

Sutherland, D. (1991). "The men went to work by the stars and came home by them": The experience of work in the Newfoundland woods in the 1930s. *Newfoundland [and Labrador] Studies, 7*(2), 143–172. https://journals.lib.unb.ca/index.php/NFLDS/article/view/1030/1382

Sutherland, D. (1992). Newfoundland loggers respond to the Great Depression. *Labour/Le Travail, 29,* 83–115. https://www.lltjournal.ca/index.php/llt/article/view/4836

Sutherland, D. (1995). *We are only loggers: Loggers and the struggle for development in Newfoundland, 1929–1959* [Doctoral dissertation, Simon Fraser University]. https://summit.sfu.ca/item/6891

Tarrant, D. R. (2008). Newfoundlanders on HMS Hood. *The Newfoundland Ancestor, 24*(4), 154–155.

Taylor, B. (n.d.). *German bombs at Taymount camp*. Newfoundland Overseas Forestry Unit 1939–1945. http://nofu.ca/submitted_items/bert_taylor_bombs.html

Taylor, M. J. (1942, March 14). [Memo to Officer-in-Charge Jack Turner]. Maurice Job Taylor Fonds (MG 390.47, Box 1). The Rooms Provincial Archives, St. John's, NL, Canada.

Temple, A. (2018). Convoy to Scotland. In U. A. Kelly & M. C. Forsyth (Eds.), *The music of our burnished axes Songs and stories of the woods workers of Newfoundland and Labrador* (pp. 345–350). Memorial University Press.

Ten little lumberjills. (n.d.). [unpublished manuscript]. Photographs and Cuttings Related to the Newfoundland Overseas Forestry Unit Collected by Tom Hall (PD 19 1). Aberdeen City and Aberdeenshire Archives, Aberdeen, Scotland.

Terms of Union of Newfoundland with Canada. (1948, December 11). https://collections.mun.ca/digital/collection/archives/id/41

Third raid on Honduras camp. (January 5, 1943). *The Berwick Journal*, 2.

30,000 men in war service. (1941, December 31). *The Daily News*, 7. https://collections.mun.ca/digital/collection/dailynews/id/118253/rec/13

Timber. (1940, February 23). Timber felling in Scotland [Letter to the editor]. *The Scotsman*, 9.

Tobacco and cigarettes for Newfoundlanders serving overseas in logging units. (1940, May 15). *The Western Star*, 5. https://collections.mun.ca/digital/collection/westernstar/id/12255/rec/6

Too many drinks—Newfoundland lumberjack fined at Lymington. (1940, July 27). *New Milton and District Advertiser and Lymington Times*, 4.

Tough men in a Highland forest. (1941, April 4). *The Evening Telegraph* (Dundee), 2.

Trade school for veterans. (1947, May). *Atlantic Guardian*, 3(5), 32–41.

Traverse, J. (1945, April). Letter to the editor. *Newfoundland Foresters' Bulletin*, 6, 2.

Travers, R. (1942, April 3). W. P. A. Lark Harbour. *The Western Star*, 9. https://collections.mun.ca/digital/collection/westernstar/id/13222/rec/1

Turner, J. (1917). Buddy's Blighty and other verses from the trenches. Imperial Publishing. https://collections.mun.ca/digital/collection/cns/id/160455/rec/1

Turner, J. (1940a, February 29). *Newfoundland Overseas Forestry Unit: Report to February 29, 1940*. Commission of Government, Natural Resources, General Administration 1939–1942 (GN 38.2, S2–1-17, File 35). The Rooms Provincial Archives, St. John's, NL, Canada.

Turner, J. (1940b, April 8). *Newfoundland Overseas Forestry Unit: Report to March 31, 1940*. Commission of Government, Natural Resources, General Administration 1939–1942 (GN 38.2, S2–1-17, File 35). The Rooms Provincial Archives, St. John's, NL, Canada.

Turner, J. (1940c, May 7). *Newfoundland Overseas Forestry Unit: Report to April 30, 1940*. Commission of Government, Natural Resources, General Administration 1939–1942 (GN 38.2, S2–1-17, File 35). The Rooms Provincial Archives, St. John's, NL, Canada.

Turner, J. (1940d, June 8). *Outline of my remarks to men in camps visited during last 10 days*. Dominions Office, Original Correspondence, 1943–46 (DO 35, Reel B 5021 /745, File N.264/52–213). The Rooms Provincial Archives, St. John's, NL, Canada.

Turner, J. (1940e, June 8). *Newfoundland Overseas Forestry Unit: Report to May 31, 1940*. Commission of Government, Natural Resources, General Administration 1939–1942 (GN 38.2, S2–1-17, File 35). The Rooms Provincial Archives, St. John's, NL, Canada.

Turner, J. (1940f, July 8). *Newfoundland Overseas Forestry Unit: Report to June 30, 1940*. Commission of Government, Natural Resources, General Administration 1939–1942 (GN 38.2, S2–1-17, File 35). The Rooms Provincial Archives, St. John's, NL, Canada.

Turner, J. (1940g, August 3). *Newfoundland Overseas Forestry Unit: Report to July 31, 1940*. Commission of Government, Natural Resources, General Administration 1939–1942 (GN 38.2, S2–1-17, File 35). The Rooms Provincial Archives, St. John's, NL, Canada.

Turner, J. (1940h, September 9). *Newfoundland Overseas Forestry Unit: Report to August 31, 1940*. Commission of Government, Natural

Resources, General Administration 1939–1942 (GN 38.2, S2–1-17, File 35). The Rooms Provincial Archives, St. John's, NL, Canada.

Turner, J. (1940i, November 7). *Newfoundland Overseas Forestry Unit: Report to October 31, 1940*. Commission of Government, Natural Resources, General Administration 1939–1942 (GN 38.2, S2–1-17, File 35). The Rooms Provincial Archives, St. John's, NL, Canada.

Turner, J. (1940j, December 4). [Letter to Sir Roy Robinson]. Dominions Office, Original Correspondence, 1943–46 (DO 35, Reel B 5021 /745, File N.264/52–213). The Rooms Provincial Archives, St. John's, NL, Canada.

Turner, J. (1943a, January 29). The Newfoundland Forestry Unit. *The Western Star*, 9. https://collections.mun.ca/digital/collection/westernstar/id/13757/rec/1

Turner, J. (1943b, August 18) [Letter to Paul Emrys-Evans, Dominions Office]. Dominions Office, Original Correspondence, 1943–46 (DO 35, Reel B 2511 /1155, File N.664/4–65). The Rooms Provincial Archives, St. John's, NL, Canada.

Turner, J. (1944, December 6). *3rd Inverness (Newfoundland Forestry) Battalion Home Guard*. Photographs and Cuttings Related to the Newfoundland Overseas Forestry Unit collected by Tom Hall (PD 19 1). Aberdeen City and Aberdeenshire Archives, Aberdeen, Scotland.

Turner, J. (1945a, March 30). [Notice]. Maurice Job Taylor Fonds (MG 390.48, Box 1)). The Rooms Provincial Archives, St. John's, NL, Canada.

Turner, J. (1945b, May 1). Annual review of officer in charge Nfld. Forestry Unit. *The Daily News*, 7. https://collections.mun.ca/digital/collection/dailynews/id/99614/rec/35

Turran, A. E. (1943). Newfoundland loggers in Great Britain. *The Geographical Magazine*, March, 528–533.

2000 lumberjacks. (1940, 22 January). *The Scotsman*, 10.

Veterans Affairs Canada—Canadian Forces Advisory Council. (2004). The origins and evolution of veterans benefits in Canada 1914–2004. https://publications.gc.ca/collections/collection_2018/acc-vac/V32-357-2004-eng.pdf

Vickers, E. (2011). 'The forgotten army of the woods': The Women's Timber Corps during the Second World War. *The Agricultural History Review, 59*(1), 101–112. https://www.jstor.org/stable/41330100

Viscount Caldecote. (1940a, June 5). [Telegram to Governor Walwyn]. Governor's Office – Local and Miscellaneous Correspondence 1936—Forestry Recruitment (GN 1/3/A, Box 187, File 118). The Rooms Provincial Archives, St. John's, NL, Canada

Viscount Caldecote. (1940b, June 20). [Telegram to Governor Walwyn and poster enclosure: To the men of the Newfoundland Forestry Unit]. Governor's Office – Local and Miscellaneous Correspondence 1936—Forestry Recruitment (GN 1/3/A, Box 187, File 118). The Rooms Provincial Archives, St. John's, NL, Canada.

Viscount Cranborne. (1941a, March 29). [Memorandum to Governor Walwyn]. Governor's Office – Local and Miscellaneous Correspondence 1941—Forestry (GN 1/3/A, Box 239, File 2). The Rooms Provincial Archives, St. John's, NL, Canada.

Viscount Cranborne. (1941b, September 22). [Telegram to Governor Walwyn]. Governor's Office – Local and Miscellaneous Correspondence 1941—Forestry (GN 1/3/A, Box 239, File 2). The Rooms Provincial Archives, St. John's, NL, Canada.

Viscount Cranborne. (1941c, December 29). [Telegram to Governor Walwyn]. Governor's Office – Local and Miscellaneous Correspondence 1941—Forestry (GN 1/3/A, Box 239, File 2). The Rooms Provincial Archives, St. John's, NL, Canada.

Vivian, A. (1949, August 5). [Letter to Premier J. R. Smallwood from VAC]. J. R. Smallwood Papers (Collection 075, Box 191, File 3.10.031). Archives and Special Collections, Memorial University of Newfoundland, St. John's, NL, Canada.

Vocational institute bids adieu to A. E. Hart. (1950, May 6). *The Evening Telegram*, 3. https://collections.mun.ca/digital/collection/telegram21/id/77183/

The Volunteer War Service Medal Act. Statutes of Newfoundland 1981. c.33, pp. 258–260 (1981, July 14). https://collections.mun.ca/digitial/collections/statutesnl/id/55626/rec/4

Volunteer War Service Medal Act. Statutes of Newfoundland and Labrador (1993). Office of the Legislative Counsel, NL: https://www.assembly.nl.ca/legislation/sr/statutes/v07-1.htm

Walsh, A. J. (1945, March 15). Government's plan for civil re-establishment. *The Evening Telegram*, 2, 5, 7. https://collections.mun.ca/digital/collection/telegram21/id/52421/rec/1

Walsh, D., & Russell, K. (1992, September 29). [Letter to the East Coast Music Association about use of the word "Newfie" in advertisements]. Don Walsh Fonds (Collection 457, 5.10.003). Archives and Special Collections, Memorial University of Newfoundland, St. John's, NL, Canada. MUN Digital Archives Initiative: https://collections.mun.ca/digital/collection/donwalsh/id/4424/rec/100

Walsh, V. S. (1918, January 2). A denial. *The Evening Telegram*, 6. https://collections.mun.ca/digital/collection/telegram19/id/12994/rec/1

Walwyn, H. T. (1939, October 3). [Telegram to Secretary of State for Dominion Affairs]. Governor's Office – Local and Miscellaneous Correspondence 1936—Forestry Recruitment (GN 1/3/A, Box 187, File 118). The Rooms Provincial Archives, St. John's, NL, Canada.

Walwyn, H. T. (1940, April 27). [Telegram to Secretary of State for Dominion Affairs]. Governor's Office – Local and Miscellaneous Correspondence 1941—Forestry (GN 1/3/A, Box 239, File 2). The Rooms Provincial Archives, St. John's, NL, Canada.

Walwyn, H. T. (1941a, September 26). [Telegram to Secretary of State for Dominion Affairs]. Governor's Office – Local and Miscellaneous Correspondence 1941—Forestry (GN 1/3/A, Box 239, File 2). The Rooms Provincial Archives, St. John's, NL, Canada.

Walwyn, H. T. (1941b, December 19). [Telegram to Secretary of State for Dominion Affairs]. Governor's Office – Local and Miscellaneous Correspondence 1941—Forestry (GN 1/3/A, Box 239, File 2). The Rooms Provincial Archives, St. John's, NL, Canada.

Walwyn, H. T. (1941c, December 31). [Memorandum to Secretary of State for Dominion Affairs]. Governor's Office – Local and Miscellaneous Correspondence 1941—Forestry (GN 1/3/A, Box 239, File 2). The Rooms Provincial Archives, St. John's, NL, Canada.

Walwyn, H. T. (1942a, January 6). [Telegram to Secretary of State for Dominion Affairs]. Governor's Office – Local and Miscellaneous Correspondence 1942—Forestry (GN 1 /3/ A, Box 245, File 2). The Rooms Provincial Archives, St. John's, NL, Canada.

Walwyn, H. T. (1942b, February 6). [Telegram from Commissioner Dunn to H. N. Tait: Copy to Secretary of State for Dominion Affairs]. Commission of Government, Natural Resources, Forestry Division (GN 38.2: S2–3-3). The Rooms Provincial Archives, St. John's, NL, Canada.

Walwyn, H. T. (1942c, July 31). [Telegram to Secretary of State for Dominion Affairs]. Governor's Office – Local and Miscellaneous Correspondence 1942—Forestry (GN 1 /3/ A, Box 245, File 2). The Rooms Provincial Archives, St. John's, NL, Canada.

Wants loggers for Labrador (1945, August 11). *The Daily News*, 3. https://collections.mun.ca/digital/collection/dailynews/id/100965/rec/3

War savings certificates—information booklet. (1940). Government of Newfoundland. https://collections.mun.ca/digital/collection/cns_tools/id/141951/rec/370

War Veteran. (1965, September 21). Foresters not veterans [Letter to the editor]. *The Evening Telegram*, 6. https://collections.mun.ca/digital/collection/telegram21/id/193979/rec/1

Waters, C. (1997). "Dark strangers" in our midst: Discourses of race and nation in Britain, 1947–1963. *Journal of British Studies, 36*(2), 207–238. https://doi.org/10.1086/386134

Watton, A. W. (1947, November 10). "Newfoundland National Convention, 10 November 1947, debates on Confederation with Canada." *1865–1949 Confederation Debates*, 695. https://hcmc.uvic.ca/confederation/en/lgNFNC_1947-11-10.html

Weather in 1940. (1941, January 31). *Jedburgh Gazette*, 3.

Webb, J. A. (2001). *Collapse of responsible government, 1929–1934.* Heritage Newfoundland and Labrador. https://www.heritage.nf.ca/articles/politics/collapse-responsible-government.php

Webb, J. A. (2008). *The Commission of Government, 1934–1949.* Heritage Newfoundland and Labrador. https://www.heritage.nf.ca/articles/politics/commission-government.php

Welfare of forestry workers. (1940, November 27). *The Scotsman*, 9.

Wigham, S. (1945, March 10). London letter. *The Western Star*, 3. https://collections.mun.ca/digital/collection/westernstar/id/15516/rec/32

Wild, I. (1942, February 20). [Letter to Hugh Tait, Dominions Office]. Dominions Office, Original Correspondence, 1943–46 (DO 35, Reel B 5021 /745, File N.264/52–213). The Rooms Provincial Archives, St. John's, NL, Canada.

Wilson, E., & Williams, J. F. (1940, August 28). Returned when contract expired [Letter to the editor]. *The Daily News*, 4. https://collections.mun.ca/digital/collection/dailynews/id/121720/rec/270

With the loggers—Sir William Brass, MP. (1944, June 2). *Clitheroe Advertiser and Times*, 2.

Wonders, W. C. (1987). The Canadian Forestry Corps in Scotland during World War II. *Scottish Geographical Journal, 103*(1), 21–31. https://doi.org/10.1080/00369228718736684

Wonders, W. C. (1991). *The "sawdust fusiliers": The Canadian Forestry Corps in the Scottish Highlands in World War Two.* Canadian Pulp and Paper Association.

Workman, W. B. (1941, August 22). [Letter to Secretary, War Office, on behalf of Officer-in- Charge J. Turner]. Dominions Office, Original Correspondence, 1943–46 (DO 35, Reel B 5021 /745, File N.264/52–213). The Rooms Provincial Archives, St. John's, NL, Canada.

Yarham, E. R. (1940, November 30). Fog and fish: Men of empire's oldest colony. *Newcastle Weekly Chronicle*, 1.

Yarham, E. R. (1944). Newfoundlanders at war and after. *Overseas: The Monthly Journal of the Overseas League, 29*(137), 14–16.

Youth hostel notes. (1940, March 29). *Daily Record and Mail*, 15.

INDEX

Page numbers in *italics* denote illustrative material; page numbers in **bold** indicate that the information can be found in the footnote.

ABOUT THE AUTHOR

Ursula Kelly is Professor Emerita at Memorial University of Newfoundland, Canada. An interdisciplinary scholar, she is the author of several books, the more recent of which focus on the history and culture of twentieth-century woods workers of Newfoundland and Labrador. *From Loggers to Foresters* completes a trilogy that began with *The Music of Our Burnished Axes* (with Meghan Forsyth), a collection and analysis of these workers' creative tradition of songs, poems, and stories. It was followed by *The Foresters' Scribe*, the story of the Newfoundland Forestry Companies, a unit of the Royal Newfoundland Regiment of the First World War, which also includes a series of letters recounting its service written during the war by Quartermaster Sergeant John A. Barrett.